Industrial AI

Jay Lee

Industrial AI

Applications with Sustainable Performance

Jay Lee
Advanced Manufacturing
University of Cincinnati
Cincinnati, OH, USA

ISBN 978-981-15-2143-0   ISBN 978-981-15-2144-7 (eBook)
https://doi.org/10.1007/978-981-15-2144-7

Jointly published with Shanghai Jiao Tong University Press
The print edition is not for sale in China Mainland. Customers from China Mainland please order the print book from: Shanghai Jiao Tong University Press.

© Shanghai Jiao Tong University Press 2020
This work is subject to copyright. All rights are reserved by the Publishers, whether the whole or part of the material is concerned, specifically the rights of translation, reprinting, reuse of illustrations, recitation, broadcasting, reproduction on microfilms or in any other physical way, and transmission or information storage and retrieval, electronic adaptation, computer software, or by similar or dissimilar methodology now known or hereafter developed.
The use of general descriptive names, registered names, trademarks, service marks, etc. in this publication does not imply, even in the absence of a specific statement, that such names are exempt from the relevant protective laws and regulations and therefore free for general use.
The publishers, the authors, and the editors are safe to assume that the advice and information in this book are believed to be true and accurate at the date of publication. Neither the publishers nor the authors or the editors give a warranty, express or implied, with respect to the material contained herein or for any errors or omissions that may have been made. The publishers remain neutral with regard to jurisdictional claims in published maps and institutional affiliations.

This Springer imprint is published by the registered company Springer Nature Singapore Pte Ltd.
The registered company address is: 152 Beach Road, #21-01/04 Gateway East, Singapore 189721, Singapore

Foreword by Terry Gou

On January 20, 2018, Dr. Jay Lee visited Foxconn Technology Group. We enjoyed discussing his research in the field of industrial big data and its applications in enterprises. In particular, I was very interested in his concept of Industrial Artificial Intelligence (Industrial AI). I had always thought that traditional AI gave people a sense of confusion and misunderstanding, but when Professor Jay Lee introduced me to his concept of Industrial AI, I found that his methodology is more suitable and applicable for current manufacturing scenarios than traditional AI. Recently, his successes of Industrial Big Data and Industrial AI have become a guiding light in transforming Foxconn Industrial Internet.

Professor Jay Lee has been involved in automation and intelligent machine technologies in the United States for almost 40 years. His earliest studies included the 3D vision guided robotics automation of automobile production lines and identifying handwriting for automated mail and parcel handling systems for the United States Postal Service (USPS). These projects have led him to accumulate rich experience. He later joined the National Science Foundation (NSF) in 1991. During his 7 years at the NSF, he was responsible for managing and funding Industry/University Cooperative Research Center (I/UCRC) and Engineering Research Centers (ERC) projects in advanced manufacturing. In 1995, Dr. Lee worked in Japan as a Science & Technology Fellow using the Internet to connect remote machines for remote data analysis, monitoring, and prediction. This work was ahead of its time and was the beginning of the concept of Industrial Internet. He joined United Technologies Research Center (UTRC) in 1998 to direct advanced product development and manufacturing. Since then, Dr. Lee has been involved in research projects across many different industries working with companies including P&W aircraft engines, Sikorsky helicopters, Carrier air conditioners, Otis elevators, international fuel cells, etc.

In 2000, Professor Lee returned to academia as a professor, continuing to promote the concepts of Industrial Big Data Analytics and Industrial AI, which at the time were very new. With his visionary forward thinking, he made a claim that was very bold at that time: "By 2010, once all parts of industry become connected, it will create the need for an analytics system to utilize this industrial data." 20 years ago, Professor Lee clearly saw the future, but at that time the hardware and software conditions were not yet mature, so he first established the Center for Intelligent Maintenance Systems (IMS). Since then, IMS has assisted more than 100 companies across 15 countries worldwide to create breakthrough innovations in Industrial Big Data and Industrial AI by facilitating cooperation and establishing professional trainings between industry and academia. According to an anonymous survey conducted by NSF, the center has a return rate of 238 to 1, which is unprecedented in history. In 2014, the IMS Center was also selected to receive the Alex Schwarkopf NSF Technological Innovation Award, which is the highest honor in the NSF I/UCRC program; and the researchers trained by the IMS Center also won first prize in nine PHM Data Challenges in the United States. These achievements have given the IMS Center the nickname of "West Point Academy" in the field of Industrial Big Data and PHM.

In July 2018, Prof. Lee officially established the world's first Industrial AI Center. The center was established in order to convey a new idea to everyone in the field of manufacturing: "When there is enough AI software and data, the focus of our research should not be on the algorithm itself, but on how to make Industrial AI work with our industrial systems." Prof. Lee has laid a solid foundation for AI's introduction to industrial manufacturing with his evident achievements and the experience accumulated during his research career of more than 40 years. I predict that the new concept and thinking of "Industrial AI" will play an important role in the future development of industrial systems.

If the Internet has already changed our lives, and the Industrial Internet is currently changing our industries, then Industrial AI will certainly change our future industrial intelligence. Different from traditional AI, Industrial AI requires cross-domain, cross-discipline, and cross-industry integrated systems engineering. Future Industrial AI must assist industrial development, growth, and transformation. Business can generate new economic benefits such as quality improvement, efficiency improvement, cost reduction, and savings reduction. In the past, we often talked about "Lean Manufacturing", "Advanced Manufacturing", "Industry 4.0", and so on, all of which are trying to solve known problems. AI uses different machine learning methods to integrate different kinds of data to mine invisible relationships and optimize systems, thus avoiding (invisible or unknown) problems that have not yet occurred. I am confident that this breakthrough in concept and technology will lead to a new revolution in intelligent industry.

This book not only leads readers to re-understand the definition of AI, but also introduces new industrial systems thinking and methods, including the 4T systems thinking (data technology, analytic technology, platform technology and operations technology) of understanding technology, commonly used AI algorithm categories, as well as new technologies and products such as Foxconn Fog AI. Its development applies virtual machine technology to onsite computers as the core method of data acquisition and decomposition, AI operations and analysis, and decision-making in manufacturing. Through the rich practical examples in the book, readers can clearly understand how Prof. Lee and his team have applied Industrial AI technology in different industrial enterprises. Foxconn Technology Group is also honored to share with readers our award-winning World Economic Forum (WEF) "Lighthouse Factory" (issued at the World Economic Forum in January 2019). I sincerely hope that the publication of this book will contribute to the future development of Industrial AI and guide the manufacturing industry towards a prosperous new age.

Wisconsin, USA
May 2019

Terry Gou

Foreword by Andrew James Hess

Industrial AI is a new emerging field which is expanding at a very fast rate. The conflux of more sensors embedded in future products, the growing availability of data, particularly digital data, smarter algorithms, and the availability of faster processors and unlimited data storage, is driving the computational intelligence that is enabling and providing the benefits associated with Industrial AI.

Prognostics and Health Management (PHM) is a set of capabilities and information products that can help enable an intelligent predictive system which can diagnose, prognose, predict, and help avoid or prevent system failures; and ultimately to achieve resilient performance across an industrial enterprise. Provided PHM capabilities include advanced diagnostics and prognostics; predictive analytics; life used, life remaining, and performance life remaining information; as well as information to better enable condition-based maintenance (CBM), asset health management, and enterprise-wide decisions. PHM is a very interdisciplinary field which involve among many things including evolving AI technologies like advanced algorithms in machine learning and deep learning. These AI technologies can be important new tools that will become part of the PHM capabilities set that will greatly benefit industry systems and their overall enterprise performance and goals. Most industrial systems necessitate consistent performance with promises uptime which necessitates AI Systems that can be reliable, repeatable, and accountable. PHM wisely integrated with the appropriate AI Systems will greatly enhance the successful implementation of any Industrial AI application.

Industrial AI is a systematic discipline to address the common issues in AI including data quality, feature extraction, algorithm selection, and platform integration (cloud, cyber, etc). It will further accelerate the development and enhancement of PHM technologies to impact broad range of industries applications.

I have witnessed Prof. Jay Lee leading the transformation of Industrial AI for the past several years, culminating the establishment of new Industrial AI Center. This book is a good summary about what his team has done for the past 20 years and where Industrial AI is going in the future.

As the president of PHM Society, I think Jay Lee and this book have provided great insights and case studies to impact the future development of Industrial Intelligence.

Washington, D.C., USA
May 2019

Andrew James Hess
President of PHM Society
President of The Hess PHM Group

Foreword by Detlef Zuehlke

Today's industries are facing new challenges in terms of market demand and competition. Strongly driven by the Internet, customers today can easily individualize their product online and then order it just by a mouse click. To stay competitive, companies have to adapt to these market requirements by opening webshops, speed up production processes for low volume production, and shorten the delivery chain. All this will be deeply driven by data. Only those industries will survive which are able to handle digitized processes and convert data to knowledge.

A part of this radical change was named Industry 4.0 as an abbreviation for the upcoming 4th industrial revolution. But Industry 4.0 was first concentrating on web-based machine connectivity and network-based factory automation. Today, we recognize that networks based on Internet standards will give us access to large amounts of data from the customers via supply chain down to millions of sensors and actuators in the machines. This "Big Data" phenomenon is a big hurdle for the manufacturing industry as they are not experienced in extensive data handling. To make them successful, we have to provide not only technologies to convert data to knowledge but also to educate them in smart data handling.

And here we can benefit from the advances in Artificial Intelligence (AI). AI is a cognitive science with rich research activities in the areas of image processing, natural language processing, robotics, machine learning, etc. Historically, machine learning and AI have been perceived as black-art techniques and there is often a lack of compelling evidence to convince industry that these techniques will work repeatedly and consistently with a return on investment. Hence, the success of AI in industrial applications was very limited. On the contrary, Industrial AI is a scientific discipline, which focuses on developing, validating, and deploying various machine learning algorithms for industrial applications with sustainable performance. It acts as a systematic methodology and discipline to provide solutions for industrial applications and function as a bridge connecting academic research outcomes in AI to industry practitioners.

Dr. Lee has proposed a 5C architecture in Industrial AI for Industry 4.0 implementation. His work has been one of the most cited papers in CIRP since 2014. This Industrial AI book has presented some practical examples which will help readers to understand how Industrial AI can be implemented in the real manufacturing world applications.

Dr.-Ing. Dr. h.c. Detlef Zuehlke
Executive Chairman SmartFactoryKL Association

Retired Director Innovative Factory Systems
German Research Center for Artificial Intelligence (DFKI)
Kaiserslautern, Germany

Foreword by Yasushi Umeda

I am very pleased to have the opportunity to write my supportive statements about Prof. Lee's new book on Industrial AI. I have known Prof. Jay Lee since 1996 when he was a Japan Society for the Promotion of Science (JSPS) Fellow at the Univ. of Tokyo. He is a pioneer in the area of Intelligent Maintenance Systems (IMS) which has made significant impacts to many world-class companies including world-renowned companies in Japan including Komatsu, Toyota, Nissan, Omron, Denso, Hitachi, etc. His earlier work on machine degradation using smart predictive algorithms has paved a way for AI in machinery health monitoring and failure prevention.

Currently, AI has been used as a black-art technique and there is often a lack of compelling evidence to convince industry that these techniques will work repeatedly and consistently with a return on investment. Professor Jay Lee proposed the Industrial AI to tackle these challenges. He also established an Industrial AI Center in US in 2018. He believes Industrial AI is a systematic methodology and discipline to provide solutions for industrial applications and function as a bridge connecting academic research outcomes in AI to industry practitioners.

I believe his new book on Industrial AI will make great impacts to our global industry. Japanese government and industry are also interested in working with Prof. Jay Lee's Team to bring about innovation in the Industrial AI as well as to nurture young talents to develop future career in this field.

Tokyo, Japan   Prof. Yasushi Umeda

Preface

In the past 10 years, many new terms and concepts have appeared: Internet, big data, artificial intelligence (AI), blockchain, Industrial Internet, etc. However, what these concepts represent and how they should be applied is not always clear. AI has been developing in the United States for more than 60 years, and while it has been successfully applied in some applications, there have been considerable difficulties in many domains. I would like to see AI demonstrate its ability in industrial systems, strengthen the industrial base, and help industrial systems transform. Over the past 40 years, I have been engaged in the production, teaching, and research of intelligent manufacturing and industrial big data in the United States. I deeply understand the challenges of talent, technology management, and execution in the process of enterprise transformation from lean to intelligent manufacturing. My motivation for writing this book is to help the reader understand what Industrial AI is and how to use it.

In 2000, I established IMS, the NSF Industry/University Cooperative Research Center (I/UCRC) Center for Intelligent Maintenance Systems, which has been engaged in the research and promotion of industrial big data and prognostics and health management (PHM). In this work, we use AI and machine learning as the basic science and tools to empower businesses, and as a result have successfully attracted more than 100 enterprises around the world, including P&G, Toyota, and General Electric, among others. In all of these companies, we have achieved remarkable results. In 2018, we also set up an Industrial AI Center in the United States, through which we hope to use AI to develop intelligent industrial systems that can be not only quickly validated, but also sustainable over time.

AI is a cognitive science, which mainly includes six fields: natural language processing, computer vision, cognition and reasoning, game theory and ethics, and machine learning and robotics. It can be applied in many fields, such as medicine, business, finance, and so on. However, a problem often found in industrial applications is that two engineers using different AI algorithms can arrive at two very different answers. This is difficult to accept in industry because, in the field of manufacturing, the three characteristics of systematic, speedy, and sustainable are needed. Unlike normal AI, Industrial AI is a method of systems engineering that emphasizes how to build a combination of data technology (DT), analytic technology (AT), platform technology (PT), and effective operations technology (OT) with the ability to solve the problems of industry effectively, repeatedly, and reliably. Traditional AI experts have strong algorithmic ability, but it is difficult to solve industrial systems problems in the absence of experts and domain knowledge. Even if a solution is found, it is often difficult to repeat. The biggest challenge of Industrial AI is to transform human-centric algorithmic thinking into systems engineering. This is a decentralized process, so that the wisdom of each industry expert can be inherited through a systematic approach.

In order to teach readers about the value of AI in industry, we divide this book into the following four chapters: why we need AI in industry, the definition and meaning of Industrial AI, "killer applications" and enabling systems of Industrial AI, and how to build the technology and capability of Industrial AI. We include case studies from the PHM Society's Data Challenges so that, through these examples, readers might better understand how to solve the problems of industrial systems by using algorithms. More importantly, readers can proceed to find the data used for these examples, verify their own algorithms, and share personal experience with other readers.

The concept of Industrial AI was first coined and introduced by me, but it will take many more people like you to extend Industrial AI to create broader impacts. I sincerely hope that this new field can be widely promoted to the broader industrial community.

Cincinnati, USA   Jay Lee
May 2019

Acknowledgements

This book is the result of the relentless efforts and contributions of many talented individuals. To begin with, I am grateful to the publisher, Springer, for the publication of this work and for assistance with formatting of the final version of the book. I would also like to acknowledge the individuals who contributed to the book, including Dr. Zongchang Liu, Dr. Hung-An Kao, Dr. Xiaodong Jia, Dr. Keyi Sun, Mr. Jack Klika, Mr. Qibo Yang, Mr. Andy Olson, and Mr. Zach Komassa. I am deeply grateful for their diligent work to make this book happen.

Contents

1	**Introduction: The Development and Application of AI Technology**	**1**
2	**Why Do We Need Industrial AI?**	**5**
	2.1 New Perspectives in Industrial Systems for AI	5
	2.2 What Are the Basic Problems in Industry?	7
	2.3 The Basic Method of Solving Problems with AI	10
	2.4 What Kind of AI Technology Is Most Suitable for Industry?	12
	2.4.1 Neural Networks: The Closest to Thinking, and Close to Solving Complex Problems	12
	2.4.2 Statistical Method: A Summarization of the Experience	15
	2.4.3 Cybernetics Approach: Systematic Design Perspectives with an Emphasis on Objects and Tasks	16
	2.4.4 Industrial AI Isn't Just Algorithms, But the Integration of People, Things, and Systems	19
	2.5 When Machine Intelligence Meets Industry	20
	2.6 Differences Between Industrial AI and AI	22
	2.7 Challenges of AI in Industry	25
	2.7.1 Reproducibility	25
	2.7.2 Data Issues	26
	2.7.3 Reliability	27
	2.7.4 Safety/Security	28
	2.8 New Opportunity Spaces for Industrial AI to Realize Industrial Value Transformation	28
	References	31
3	**Definition and Meaning of Industrial AI**	**33**
	3.1 The Beginnings of Industrial AI	33
	3.2 The Purpose and Value of Industrial AI	39
	3.3 GE Predix Successes and Failures	44

xix

| 3.4 | Technical Elements of Industrial AI: Data, Analytics, Platform, Operations, and Human-Machine Technologies | 48 |

3.4 Technical Elements of Industrial AI: Data, Analytics, Platform,
Operations, and Human-Machine Technologies 48
3.5 CPS: An Architecture for Integrating the 5 Technological
Elements of Industrial Intelligence 52
3.6 Industrial AI: Categories of Algorithms 54
    3.6.1 Regression Algorithms 56
    3.6.2 Classification Algorithms 56
    3.6.3 Clustering Algorithms 57
    3.6.4 Statistical Estimation Algorithms 57
3.7 Industrial AI Algorithms: Selection and Application 58
References 61

4 Killer Applications of Industrial AI 63
4.1 Application Scenario Types for Industrial AI 63
4.2 What Will Become the "Killer Applications" of Industrial AI? ... 65
    4.2.1 Predictive Maintenance of Equipment 65
    4.2.2 Virtual Metrology and Process Quality Control 74
    4.2.3 Energy Management and Energy Efficiency
          Optimization 81
    4.2.4 Defect Detection and Material Sorting Based on Machine
          Vision 90
    4.2.5 Scheduling Optimization of Production and Maintenance
          Plans 94
4.3 Enabling Industrial AI Systems 102
    4.3.1 Intelligent Monitoring and Maintenance Platform for
          CNC Machines 102
    4.3.2 Intelligent Operations and Maintenance System for
          Offshore Wind Farms 106
    4.3.3 Intelligent Rail Transit Predictive Maintenance System ... 112
References 116

5 How to Establish Industrial AI Technology and Capability 119
5.1 Assessment of Basic Capability Maturity During Industrial
Intelligence Transformation 119
5.2 Assessment Tools for Global Industrial AI Enterprise
Transformation Achievements 124
5.3 Foxconn Lighthouse Factory 129
5.4 How to Construct the Organizational Intelligent Transformation
Ability in Industrial Enterprises 133
5.5 Open Source Industrial Big Data Competitions 137
References 158

6 Conclusion 161

Chapter 1
Introduction: The Development and Application of AI Technology

2017 could be called the first year of the reign of artificial intelligence (AI), after which everything has accelerated. AI breakthroughs are endlessly emerging, such as the diagnosis of diseases by medical image analysis, self-driving cars, facial-recognition payments, and cashier-less supermarkets. AI has disrupted business, becoming a new arms race for enterprises. In 2017, mergers, acquisitions, and investments in the field of AI reached $22 billion USD: 26 times as much as before Google AlphaGo defeated top Go players Le Sedol and Ke Jie in 2016. According to Google CEO Sundar Pichai, AI is probably the most important thing humanity has ever worked on. I think of it as something more profound than electricity or fire. The McKinsey Global Institute gives a more specific value, saying they can generate $2.7 trillion in economic value over the next 20 years by applying AI to marketing, supply chain management, and increasing the efficiency of new sales methods.

AI technology did not become important or significant overnight; it began changing our lives and work even in the 20th century. In the early 1980, I explored how to make robots on production lines more precise. When I worked for the United States Postal Service (USPS) in 1988, I led the first project to use machine vision and handwriting analysis technology to automatically sort packages, that was on the cover of AI Magazine. In that era, AI could only accomplish the well-defined tasks that often have restrictions on working conditions, objects, and the environment. The advantages of the generalization and advanced cognitive ability of today's AI did not exist. Today's explosive development in AI is mainly due to the continuous improvement of infrastructure technologies: high performance computing clusters make training AI much faster, sensors are becoming cheaper and more diverse creating more opportunities for data collection, more mature software provides a faster pipeline for the flow and management of data, and the huge amount and variety of data generated every minute provides more opportunities to train algorithms. More powerful chips allow these intelligent algorithms to process and analyze data in real time on edge hardware, then transmit information to people to complete various tasks. This continuous improvement of IT infrastructure reduces the cost of using AI dramatically, just as improvements in electricity infrastructure have reduced lighting costs by a scale

© Shanghai Jiao Tong University Press 2020
J. Lee, *Industrial AI*,
https://doi.org/10.1007/978-981-15-2144-7_1

of hundreds since the 1800s. AI has become usable in a wider range of scenarios with a lower barrier to entry.

The impact of AI on industrial production and our way of life has already exceeded our wildest expectations. Ubiquitous cameras record our every move. When entering a store, salespeople can quickly obtain information about us using facial recognition and recommend products we might be interested in. At Amazon, wearable sensors track employees' hand movements, identify their efficiency, and vibrate to warn inefficient work. At a semiconductor manufacturing plant in China, managers can use trackers in employee name badges to identify abnormal activities, which is a useful method to prevent technology leakage at companies with high confidentiality requirements. Slack, the most popular office messaging and collaboration tool in the US can help managers assess employee efficiency and effectiveness while interacting with teams. This made more sense after Slack's CEO Stewart Butterfield shared that the company name is an acronym for "Searchable log of all conversation and knowledge." The basis of the application is the data it collects from communications between employees. This data can be analyzed to find which employees are most efficient, and which are most likely to leave.

In business activities, AI helps companies better understand their customers and make more accurate marketing decisions. Ocado, an English online shopping company, uses AI to prioritize responses to over 10,000 emails received from customers every day. Their technology can classify customer complaints and forwards them to relevant service personnel. Caesar's, a well-known casino group, reduced the number of calls to their service desk by 30% by providing virtual concierge service in their hotels to automatically answer customer queries. There is an American start-up company that provides emotional recognition services for insurance companies' telephone customer service representatives by identifying the customer's mood through analysis of the speed, intonation and word characteristics of their voice. It then reminds representatives to show more sympathy at the appropriate time. Leroy Merlin, a French home decor retailer, has reduced inventory by 8% and increased sales by 2% by analyzing the relationship between historical sales data and other factors like weather to provide advice on order strategy and supply chain management.

Applying AI to enterprises in this way to forecast demand and improve inventory can improve cash flow and free up storage space. In today's era of decreased retail profit margins, this is good news. A simple reduction of 3% to supply chain costs could double net profit. According to Integrated Hydraulics Limited (IHL) Group data, global retail losses due to inventory backlog reached $470 billion in 2015, while losses due to insufficient inventory or turnover reached $630 billion. The application of AI in supply chain optimization may also be a growing, trillion-dollar market.

AI has experienced several ups and downs in the course of its development. Since the founding of AI as a field of research at the 1956 Dartmouth Workshop until the year 2000, AI experienced two "AI winters." The first began in 1973, represented by the publishing of the Lighthill Report. Because of the gap between anticipated AI progress and real results, the US government cut funds to AI research. In the following decade, AI was hardly mentioned; but in the 1980s, because of the emergence of expert systems and Bayesian theory in the field of AI, the winter ended, and a new era was ushered in.

However, people began to realize that there were bigger challenges in software and algorithms than hardware. With the commercial failure of the Lisp machine in 1987, AI again slipped into a downturn. In the late 1990s, with the continuous improvement in computational power, AI again began to grow in influence and significance. Representative events include IBM's Deep Blue defeating chess champions and business intelligence applications beginning to provide value to enterprises. The origins of this early climax in AI in large part to such successful breakthroughs in deep learning applications.

In 2012, the Geoffrey Hinton team used deep learning technology on ImageNet for the first time to outperform other teams, making people aware of the advantages of deep learning over traditional machine learning, bringing deep learning to the forefront for the first time. At the same time, graphics processing unit (GPU) replaced central processing unit (CPU) in computer vision training, shortening the length of training from several months to a few days or hours. Breakthroughs in hardware, algorithms, and big data in various fields brought an explosion in AI growth. These three breakthroughs came together when Google's AlphaGo beat Go master Le Sedol—Thus, Google's powerful machine learning hardware, years of experience creating algorithms, and power data infrastructure brought AI into the public consciousness, creating another climax.

Unlike the previous two climaxes, this time we see AI technology being widely used in industries like smart security, smart finance, and smart retail, on top of bringing convenience to people's lives. More importantly, AI has proved its commercial value in some areas, formed standards, and made profits. This application to business makes it fundamentally different than the past two peaks in AI technology. We have reason to believe that this new round of development shows AI in a golden age of growth. Deep learning is representative of the maturation of advanced industry applications, giving us tools for solving industrial problems.

The change of AI from the domain of academic-driven to a commercial-driven development model doesn't only provide guidance for solving urban, social, economic, and ecological problems, but also provides solutions for fields like industrial manufacturing, energy, electricity, healthcare, and transportation.

The rapid development of AI in recent years mainly benefits from the four following factors:

1. Interconnection between humans and machines creates explosive growth in data, forming a big data environment.
2. Cloud computing, edge computing, and proprietary chip technology create significant growth in computing power.
3. Breakthroughs in deep learning drive continuous optimization of algorithmic models.
4. The deep coupling of capital and technology helps promote the rapid rise in industrial applications and technology industrialization.

Driven by data, computing power, algorithmic models, and multiple applications, the application scope of AI has gradually evolved from "narrow AI" to "general AI" which is competent for open problems such as human-computer dialogue. With closer integration of AI technology into human life, the relationship between human beings and technology is moving from supplementary technology to AI becoming a partner or assistant. We are seeing the evolution of AI technology from single-threaded CPUs to multicore GPUs, and now to optimized-architecture tensor processing units (TPUs). There is an evolution from closed source frameworks to open source frameworks, from local knowledge to knowledge bases, from single-machine networks to distributed networks, and from solely academic research to rapid iteration in commercial applications.

We are delighted to see AI technology currently manifesting a positive cycle through the steps of "Scenario—Data—Technology—Product—Business." In the early days of AI technology development, because of the high cost of data acquisition and lack of a solid AI ecosystem, the algorithms lacked the environment needed for rapid evolution. As it enters mainstream usage, higher quality data and faster feedback lets us analyze technical drawbacks. Upgrades are improving the applications of AI technology, and users continue to create larger amounts of data which further improves the technology.

It is difficult to make a judgment as to whether this is the prelude or climax of AI technology, but one thing is clear: AI has already changed our lives. At the same time, however, it faces challenges that cannot be ignored. We can already see that AI's application has had a practical impact and provided value in many fields. These applications are the foundation for bringing AI to the industrial field and are contributing to AI's future potential.

Currently, industry is shifting towards the phase of Industry 4.0. With value creation as the goal and driving force, it provides suitable soil for the development and industrial application of AI. The development of intelligent manufacturing and Industrial Internet technologies has allowed us to see a trend in which the "localized intelligence" and "interconnected intelligence" directions of industrial systems go hand in hand, and are evolving towards the integration of horizontal and vertical intelligent application systems. On one hand, the rapid development of sensors and communication technologies allows for a big data ecosystem and powerful computing power for industrial systems. On the other hand, the development of Industrial Internet makes for more connected equipment, people, and services, and generates more value for businesses. The gradual maturity of these elements leads us to believe that industry will be a new territory for the development of AI technologies.

Chapter 2
Why Do We Need Industrial AI?

2.1 New Perspectives in Industrial Systems for AI

Terry Gou, Founder of Foxconn, and I gave a speech about Industrial AI at Stanford University on July 5, 2018. Many students, faculty, and AI scientists from Silicon Valley attended the event. One AI expert expressed their concern that they were not providing as much value as his high salary indicated. What would he do if the bubble burst? It was then that I realized the anxieties about AI being a sort of bubble was a global concern. There are a lot of expectations about the value that AI will provide, and the market is paying for these expectations rather than exact value. The biggest challenge facing AI right now is finding more applications which can provide value to meet the growing expectations of the market and capital.

AI has solved problems and created commercial value in many areas such as e-commerce, financial technology, security, and mobile payment; however, AI in industrial systems is still limited. AI scientists say this is because AI concepts and technologies are still in a stage of infancy, there is a lack of data, and there is a need for high performance computing. At Foxconn, we have met many AI companies interested in industrial scenarios. I've noticed that they first ask us what data we have before asking us what problems we want to solve. I think this is because the two perspectives are different: in the past, AI applications looked for hidden relationships in data, whereas industry needs to begin with problems and create value by creating solutions.

Besides differences in thinking between humans and AI, AI technology has other limitations and bottlenecks. We need to understand what AI can and cannot do, so we can establish reasonable expectations. We need to let AI do what it is good at, and at the same time respect and support fundamental, traditional technologies. More importantly, we need to understand that AI cannot help people make breakthroughs in fundamental, basic knowledge such as scientific principles, technological capabilities, designing and manufacturing capabilities, system engineering, basic materials, and experimental verification. This basic knowledge is the foundation of industrial domain knowledge, so while we can expect AI to optimize and enhance business

© Shanghai Jiao Tong University Press 2020
J. Lee, *Industrial AI*,
https://doi.org/10.1007/978-981-15-2144-7_2

value, forecast and avoid problems through modeling analysis, and reduce labor on the basis of existing fundamental knowledge, it cannot help us break through existing knowledge.

AI can discover opportunities to continuous optimize efficiency. When integrating AI with industry, we need to explore more specific application scenarios, design systematic technical frameworks, formulate technical standards that match industrial systems, and make sure it all fits with existing business processes.

Another limitation of AI technology is that problems can be resolved in many ways probabilistically rather than deterministically. This limits AI's application to be more opportunity-oriented rather than problem-oriented. When data scientists enter factories, they first care about what data they have, not about what problems to solve. When factory managers ask them what value they can bring, they want to build big data environments and powerful computing infrastructure first, then look for opportunities to create value from data.

In the e-commerce sector, possibility-oriented applications bring some commercial success, such as finding products consumers may be interested in based on historical consumption, videos that the customer may want to watch based on prior viewing patterns, or Google more accurately targeting ads by analyzing the user's preferences. These applications are opportunity-oriented, seeking rapid response and emotional impact. Accuracy, certainty, and deterministic behavior are not important, but it is important not to miss opportunities that can generate business value.

The biggest challenge of AI is not the technology, but how to produce application value. In the past, more attention has been paid to social and business problems, which are more divergent, opportunity-oriented perspectives. However, this is not a problem of the AI technology, but how to use it. It's comparable to how many people use Excel but do not understand financial statistics, cannot analyze the stock market, or cannot manage enterprises. Mastering an AI algorithm does not mean someone can solve a problem. AI education today mostly teaches theory and tools, but is missing teaching how to solve problems.

Industrial problems can be diminished using AI, such as the identification and prediction of quality problems. There can't possibly be thousands of quality problems in a factory, so the problem domain is limited. Solving core problems can bring breakthroughs.

Currently, 40% of AI startup investment failures are due to lack of applicability, 30% are due to financial problems, 20% are due to divergence of founding teams, and 10% are due to lack of competence and technology. The key factor to success is whether we can find the right application field. We should judge a company's technology to make sure it's valuable and ensure it will solve our problems.

Many AI technologies focus on problems that have already been solved and replace the original solutions.

We also find that some AI application scenarios in Chinese enterprises are misleading, including some applications borrowed from the United States (US) that were misrepresented. For example, Amazon launched their "Just Walk Out" supermarket which puts forward the idea of no sense of waiting for payment, but in China this was mainly seen as an unmanned supermarket. This caused the value of this application to be misrepresented as "no people."

In the industrial sector, there are a variety of domains and a very clear idea of what needs to be improved. It's well known that costs need to be reduced, inventory should be minimized, battery efficiency should be increased, and wind power output be maximized. There are so many demands and opportunities, so why not change the direction and apply AI technology to industry? The usage of AI in China cannot ignore the most basic needs and cannot be hyped for purposes of investment. There are basic problems to be solved!

In conclusion, the new opportunities and perspectives provided to the industrial sector by AI technology include the following:

- Industry provides AI with a new perspective on value, moving from opportunity-oriented divergent applications to the problem-oriented convergent applications that arrive at conclusions and answers.
- The application of AI needs to focus more on solving problems that have not been solved in the past rather than creating new needs or looking for alternative solutions to problems that has been already solved.
- Within the industrial sector, problems are more concrete and have more clear value standards. Solving problems in industry are new opportunities in AI technology applications.

2.2 What Are the Basic Problems in Industry?

Within industrial systems, many exact problems must be solved. These include equipment health, product quality, overall equipment effectiveness (OEE), process parameters, comprehensive costs, etc. They need quantifiable and exact returns on value rather than expecting a possibility. This reproducibility is the biggest challenge for AI and big data in industry. Compared to the upgrading of equipment and improvement of processes and improvement of organizational culture brought by lean management, AI has no definite value proposition. Even if General Electric (GE) puts forward some power towards promoting the value of the Industrial Internet, gaining value is just a possibility. AI cannot start from the possibility of value, but should focus more on the certainty of solving problems.

We must solve the shortcomings and basic problems in the industrial system to achieve breakthroughs in knowledge and ability and to enhance the efficiency and value of the system. We will focus on solving three problems (see Fig. 2.1) in manufacturing quality: **discipline problems**, **system problems**, and **intrinsic problems**.

- **Discipline problems** are related to worker ability, organizational culture, and management ability. Japan is the leading country in this regard, with a specialty of skilled workers. Their method of knowledge transfer creates a strong organizational culture.

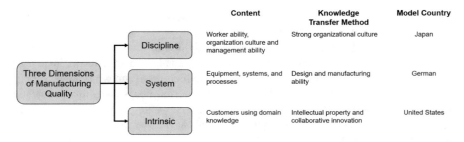

Fig. 2.1 Three dimensions of manufacturing quality

- **System problems** are within equipment, systems and processes. This is something that Germany has always handled well, relying on well-crafted equipment. Their method of knowledge transfer are the process standards and the capabilities of equipment design and manufacturing.
- **Intrinsic Problems** are related to problems in customer value creation. Solving this is an American strength, and relies on innovation of business model and implementation of technology. Their method of knowledge transfer is collaborative innovation based on intellectual property, using domain data and continuous service innovation.

Industrial Artificial Intelligence (Industrial AI) brings opportunities in manufacturing, which should consist of rapidly improving the quality, constitution, and essence in manufacturing. For example, by standardizing the workflow of a person with data and establishing a better reference and relationship with data, people's experience can quickly be accumulated and passed on. By using data, implicit problems in manufacturing systems can be explicit, so that the health status of equipment can be transparently managed, process parameters can become more stable, and comprehensive efficiency can be more coordinated and optimized. Data is used as a medium to increase users' value, to help users enhance the functionality and reliability of products and equipment, optimize operating efficiency, and enhance the sustainable profitability of enterprises. These are the problems the industry needs AI to solve.

As AI enters industrial systems, it must fulfill five principles. We will call these the 5S's:

- **Systematic**: A systematic architecture which integrates related technologies, including a standard interface and system of systems (SoS) that integrates industrial AI and industrial systems.
- **Speed**: Speedy building systems, modeling, validation, and deployment. This includes solving fragmentation problems and rapidly meeting customer's urgent needs.
- **Streamline**: Focusing on problem-oriented convergent processes, managing processes to ensure the certainty of results, and integrating with existing industrial system processes.

2.2 What Are the Basic Problems in Industry?

- **Standards**: Establishing standardized data, analysis, formats, and operations on the foundation of existing industrial systems.
- **Sustainable**: Allowing repeated successes, and consistency of models built by different developers or consistent performance of different objects. This includes managing uncertainties.

Therefore, Industrial AI is a kind of systems discipline and requires the integration of several technical elements, the six of which we will refer to as "ABCDEF:"

Analytics[1]: Not only the algorithm, but the application modeling of the algorithm in specific scenarios and for certain objectives.

Big Data: Unlike focusing on the "4V" (volume, velocity, variety, veracity) characteristics of Internet big data, Industrial big data technology needs to manage the "3B" challenges[2]—namely Broken data (data integrity), Bad quality data, and Background (context and domain of data).

Computational Platform: Encompasses cloud computing, edge computing, and fog computing based on embedded intelligence. It needs to integrate endpoints with the cloud, and integrate distributed systems with centralized ones, allowing quick reconfiguration of the platform.

Domain Knowledge: It's necessary to have basic domain knowledge and engineering experience about the mechanisms, process flow, system engineering, and optimization objectives of the application objects.

Evidence: Evidence that reflects the current state of the system and the insights that support decision making.

Feedback: Integration with control systems or standard operating procedures (SOPs) to achieve a closed loop from decision-making to operations.

Among the six technical elements listed above, "ABC" are also the three elements of AI technology (algorithms, big data, and computing power), but are different than Industrial Intelligence. Regarding analysis and modeling, analysis focuses on algorithms, while modeling focuses on scenarios and problems. In data technology, the former focuses on solving "4V" challenges, while the latter manages "3B". On computing platforms, the former focuses on cloud computing and centralized computing capabilities, while the latter focuses on the integration of endpoints to the cloud. "DEF" are technical elements with distinct industrial characteristics, and also an interface between AI and industrial systems.

AI technology for industrial systems should be integrated and empowering, not subversive. The goal of the future Industrial Intelligence system is to create a worry-free industrial environment to achieve zero accidents, zero pollution, zero waste, zero defects, and zero downtime.

---

[1] This could also be "Algorithms," but "Analytics" is more general, encompassing algorithms as well as other aspects.

[2] See the book "Industrial Big Data: The Revolutionary Transformation and Value Creation in INDUSTRY 4.0 Era" [1] for an introduction to the differences between Industrial Big Data and Internet Big Data.

2.3 The Basic Method of Solving Problems with AI

Before we begin to discuss AI, we should first have a clear and general understanding of AI technology, such as how it solves problem and the way they perceive the world.

In Fig. 2.2, AI technology can be divided into seven categories according to the way to solve problems: detection, clustering, classification, prediction, optimization, generalization ability, and creation ability. The first five categories were researched extensively in classical machine learning before 2006. After the emergence of deep learning technology, applications with the ability of the latter two have gradually emerged, which could be considered the beginning of strong AI. The seven categories of problems to be solved by AI can be arranged in the order shown in Fig. 2.2. The realization of the former problem is often the premise of the latter.

For example, the accessibility of optimization problems is often based on good prediction models, so that the objective can be predicted when the decision variables are constantly adjusted, and will gradually converge to the optimal objective. Almost all the algorithms for solving programming or optimization problems differ greatly from the convergence path or search method to the optimal equation, and their performance depends largely on the accuracy of the prediction model or mathematical model.

Detection: The process of data processing and feature extraction based on prior knowledge and analysis objectives. Detection is also an abstract process for analyzing objects, to extract the most critical and effective small amount of information that could represent the whole situation. This is how we look at the world—our vision has focus. This focus is placed where information collection efficiency is the highest, and you do not always need to look at the whole scope or every bit of information to judge it. Sometimes only a few key features can complete the judgement. Our brains evolved over time to use this method as the most efficient way to process information. Taking the example of fault detection in industrial scenarios, when we want to detect a fault on an outer race of a bearing, we need to focus on extracting features of specific frequencies that can distinguish it from normal bearings. When analyzing faults in an inner race, we would look at other frequencies. Before solving a detection problem, people need to define it by using domain knowledge at first. But the dependence on domain knowledge itself is an issue because there is something that people may not know. There is a quote from US Secretary of Defense Donald Rumsfeld that illustrates this point: "There are known knowns; there are things we know we know.

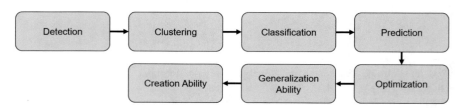

Fig. 2.2 Seven categories of AI algorithms

2.3 The Basic Method of Solving Problems …

We also know there are known unknowns; that is to say we know there are some things we do not know. But there are also unknown unknowns—the ones we don't know we don't know."

Clustering: The best way to aggregate an index in a specific range of samples. The most commonly used indicators are distance and likelihood values. The aim is to maximize the similarity of samples in the same cluster, and the difference between samples in different clusters. Clustering is the most common method of unsupervised learning and is still an open problem. Different variables, different definitions of indicators, different numbers of aggregates, and different initial conditions will affect the results of clustering. Clustering is also the most commonly used tool in large data mining. In many applications of business intelligence, analysts use the characteristics of various dimensions in the data to aggregate customers, then go to find the common features in each cluster as to create more suitable and targeted business strategies.

Classification: Contrary to problems solved by clustering, if we have an existing aggregation or labels of samples, the best tool to train would be classification algorithms. Classification is a typical supervised learning and convergent clustering problem: taking a new sample and finding a group it should be in.

Prediction: Establishing a mapping relationship between a group of explicit variables and a group of implicit variables, or using measurable objects to predict unmeasurable objects, and using current observations to predict future states. When the hidden variable is discrete, prediction is a lot like the pattern recognition and classification. When the hidden variable is a continuous number, the prediction problem can be treated as regression.

Optimization: The process of optimizing objectives by adjusting decision variables under certain constraints. This problem can be divided into two processes: mathematical modeling and solving. Mathematical modeling includes quantitative definition of objectives, constraints, and mapping relationships from decision variables to objectives. Because of this, predictability is what is planned, and the accuracy of prediction is the guarantee of finding the optimal solution. Mathematical solving can be understood as a search framework, as in the process in which the decision variables converge gradually to the optimal objective.

We use AI examples to further illustrate the nature of these categories. Take facial recognition for example. The process includes face detection and calibration, face feature extraction, face clustering and facial recognition.

Face Detection and Calibration: Detects the location of the face in the picture. The input of the face detection algorithm is a picture of a face, and the output is one or more "face frames"—bounding boxes on the picture which shows where the algorithm believes a face to be. Generally, this output face frame is an upright rectangle, but newer algorithms allow for parallelograms or rotated rectangles.

After the location of the face is detected, image registration is needed to locate the coordinates of key facial features on the face. The input of the registration algorithm is a face picture and a face frame, and the coordinates of the nose, eyes, ears, and mouth can be output.

Face Feature Extraction: The process of transforming a face image into a series of fixed length values. These values are called "Face Features" and have the ability to represent the face numerically. The input of the facial feature extraction process is a face image and coordinates of facial features, and the output is a set of values corresponding to the unique features of the face. This allows a data type to represent a unique face, allowing for efficient facial recognition. Recently, deep learning methods have dominated facial feature extraction algorithms, which are all fixed-time algorithms. Previous facial feature models were large and slow, which have few real-time applications; however, the latest research unveils a proprietary chip which optimizes chip architectures for specific applications and can optimize the size and speed of the model.

Face Clustering: An algorithm for grouping faces in a set according to their identities. Facial clustering also compares all faces in the set with each other, analyzes similarity values, and divides people that belong in certain categories into a cluster. Before labeling, only one group of faces belongs to the same identity, but the exact identity is not known.

Facial Recognition: An algorithm that takes facial features as an input and identifies the corresponding identities in a database, finding the identity with the highest similarity. If there is no identity that has a suitable amount of similarity with the input picture, a "not in library" error is returned.

Understanding what problems machine learning needs to solve and how to solve them is of great significance for us to recognize the abilities of machine learning, and to correctly understand and define the problems to be solved as to meet our goals. In the application of Industrial Intelligence, we face many of these problems and should understand the algorithms listed above.

2.4 What Kind of AI Technology Is Most Suitable for Industry?

In the history of AI, it has gradually been divided into several perspectives of thought. These perspectives have their own methodologies, understanding of tasks, use of technical tools, and philosophy of intelligence. Each perspective wants to verify the universality of their theories by solving every problem. Which of these perspectives will be more suitable for solving problems in industry? What are their advantages and disadvantages? In this section, we propose open discussions and provide suggestions for the selection of technology and optimization direction of various algorithms.

2.4.1 Neural Networks: The Closest to Thinking, and Close to Solving Complex Problems

Many bionicists (those studying biologically inspired engineering) use evolution, neuroscience, and animal behavior to create incredible algorithms. In April of 2000,

2.4 What Kind of AI Technology …

a team of neuroscientists at Massachusetts Institute of Technology (MIT) rewired a ferret's brain, changing the connections between the eyes and auditory cortex, and between the ears and visual cortex. The results published in Nature [2] shows that the auditory cortex learned to see, and the visual cortex learned to listen. This demonstrates that the functional cortex of the brain may not be specialized, but can be reorganized and trained to acquire new abilities according to needs.

All information in the brain exists in the form of neural electrical signals. Memory is formed by strengthening these connections between neurons discharged in clusters, which is a biological process called long-term potentiation. The nervous system is responsible for everything we can see and imagine. If something exists and the brain can't learn from it, we don't know it exists. From the earliest multilayer perceptron to current innovations in deep learning architecture, these networks imitate the memory, learning, and processing information from the human nervous system. The design follows two basic principles: information is stored in the network through neuron activation states and information distributed through the network, and information is processed by the interconnected network of neurons.

The invention of a neural network is an important breakthrough in intelligent algorithms because it solves two important problems: approaching arbitrary mathematical functions after enough iterations without manually constructing mathematical formulas and improving computational efficiency through parallel computing.

Biodiversity comes from one rule: competitive choice. Whether this competition mechanism comes from natural selection or social selection, or when the selection criteria are deliberately or unintentionally determined, various forces will work together to accomplish a goal. Computer scientists are very familiar with this mechanism: we try alternative methods to solve problems and select and improve the best solutions.

Many algorithms use the concept of competitive learning. For example, in the training process of a machine learning model, the neural network determines which neurons will be activated or suppressed by numbering the labels according to the training data. In this way, the "pattern" of the training data can be recorded. Reinforcement learning, by defining a value function, gradually completes a complex task in an iterative manner. For example, Q-learning in reinforcement training scored the next action with $Q(s, a)$ in the given state s, and then selected the highest scorer as the optimal strategy.

Such ideas are used in optimization algorithms such as genetic algorithms, particle swarm optimization, ant colony algorithms, etc. Unlike competitive learning where the "winner is king," these optimization algorithms mostly adopt the method of "cooperative competition" where in each iteration, all individuals aggregate towards the best result.

AI is developing rapidly, especially in deep learning. In April of 2018, McKinsey Global Institute released a report called "Notes from the AI Frontier: Insights from Hundreds of Use Cases" [3] (see Fig. 2.3 for common AI algorithms). It explores more than 400 applications across 19 industries, 9 business functions and their economy potential. The AI technology in the report refers to the use of artificial neural networks

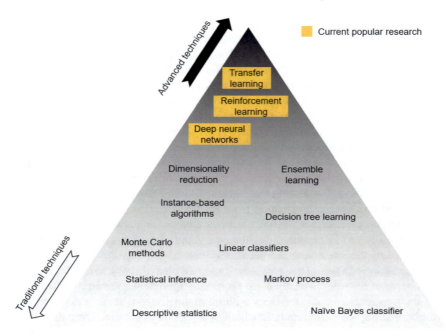

Fig. 2.3 Analysis techniques: the classics and frontier of AI algorithms [3]

and other traditional machine learning methods to replace the work done by others (Fig. 2.3).

Specifically, there are three main types of neural networks: feedforward neural networks (FFNN), recurrent neural networks (RNN), and convolutional neural networks (CNN). Other common machine learning algorithms include decision tree algorithms (using tree structure to build decision models), regression algorithms (predicting continuous value), classification algorithms (predicting discrete value, knowing classification beforehand), clustering algorithms (predicting discrete values without knowing classification beforehand), and ensemble learning (integrating several learning models). Other machine learning technologies are divided into two important technologies: generative adversarial networks (GANs) and reinforcement learning.

Although deep learning has made amazing achievements in image processing and semantic recognition, there is a long way to go until it has wide adoption in industrial systems. The most important reason is because there are too many surprises in the results and operation of neural networks—they can't explain how they came to their results. Can we really depend on them?

Although some achievements have been made in the research on explaining tree-based rules from deep networks, visualizations of convolutional layers, feature extraction, and InterpretNet, it will take a long time until we can predict the

2.4 What Kind of AI Technology …   15

results of machine learning. Applications in the industrial domain have strict requirements for accuracy and need to explain reasonableness of prediction results. Only by understanding the causes of uncertainty can it be managed.

2.4.2 Statistical Method: A Summarization of the Experience

From statistic perspective, all forms of learning are based on a very simple theorem: Bayes' theorem. It mathematically explains how to update ideas as you see new evidence, and transforms data into knowledge or experience. According to the theorem, there is no completely correct understanding of the world—understanding depends on experience accumulated from observation.

Bayesian methods have been applied to other fields as well, including linguistics. Fred Jelinek, a researcher of linguistics and information theory, introduced a new perspective in natural language processing by theorizing that speech recognition is to infer the signal sequence and meaning of the speaker according to the received signal sequence. He assumed that every character was only related to previous characters, and that sentences could be represented by Markov chains. Many linguists have questioned this model's effectiveness, but it proved to be more effective than any other for its time. Microsoft used this model in the 1980s to develop their speech recognition system.

These are vivid examples of the essence of Bayes' theorem: when you cannot accurately infer the attributes of an object, the more events that support an attribute makes it likelier to be important. Unlike traditional statistical methods, Bayes' Theorem allows people to make initial guesses based on prior knowledge and experience. People can first make assumptions based on their prior knowledge, and then revise them to approach objective facts through continuous observation.

Bayesian learning put more emphasize to the management of uncertainty and using experience to structure the problem. This puts Bayes' theorem and its derivative algorithms in line with the needs of industry because all knowledge is uncertain, and the process of learning knowledge is also an uncertain form of reasoning. The key to learning is recognizing similarities and deriving other similarities. Simple statistical methods have been widely used in industry since the 1970s, including process control, quality management, machine thermal compensation modeling, error flow analysis in manufacturing design, virtual measurement in semiconductors, and so on.

Bayesian statistics has some advantages over classical statistics. For example, classical statistics is better suited for simple problems where there is enough data that represents the characteristics of the data. When dealing with problems with many parameters, it can be handy; however, if we can only grasp a small amount of the data relative to the problem's complexity, Bayes' theorem would work much better due to data scarcity.

But today, in the era of big data, is lack of data still a problem? It isn't a problem with few parameters, but in applications like gene sequencing where 100 genes are involved in the occurrence of a certain cancer (a conservative estimate considering each human has around 20,000 genes), there are two raised to the two hundredth power combinations. According to the sampling theorem, classical statistical methods need to collect 1–10% of samples to determine the cause of a disease, so trillions of people with the disease need to be generated.

For sample scarcity in the industrial sector, some manufacturing systems have reached 6-sigma, where there are only 3.4 defective products per million. If our goal is to find the correlation between parameters and final quality and predict the quality from the process's parameters, the samples are very few. Classical statistical methods cannot explain the phenomena caused by complex and interrelated parameters. We've solved the simple problems, and all that is left are the complex ones. To reveal the laws behind these problems, we must understand their causes and networks, using human experience and scientific principles to work them out. But as problems become more and more complex, the network becomes more complex, so there is an exponentially increasing demand for computing power and data. This is also the issue with Bayesian methods compared to neural networks, as they aren't as efficient in dealing with large and complex problems.

2.4.3   Cybernetics Approach: Systematic Design Perspectives with an Emphasis on Objects and Tasks

In 1948, an American mathematician named Nobert Winer first used the word "Cybernetics," which can be traced back to the Greek "kybernetes" which means "helmsmen" or "navigator." Wiener regards cybernetics as a science which studies the communication and control between machines, animals, and society, how systems maintain equilibrium and stability under changing environmental conditions, and the "art of self-governance" as seen in Plato's *The Alcibiades*.

In cybernetics, "control" is defined as the need to acquire and use information to improve the function or development of a controlled object. If AI algorithms are used in information processing in order to perform self-learning and manage uncertainty in an object, it could be called an intelligent control system. Therefore, the purpose of the control systems is control, not intelligence. These intelligent control systems consist of at least four elements in a closed-loop system: communication loops, control loops, adaptive loops, and learning loops.

Common intelligent control algorithms include fuzzy control, evolutionary control, genetic control, multi-agent control, fuzzy neural networks, and so on. Intelligent control helps us cope with the complexity of tasks, which includes three aspects: object complexity (varying time, non-linear characteristics, randomness), environment complexity (unclear or undetected parameters), and objective complexity (adaptiveness, robustness, fault tolerance, multi-objective optimization). In a variety of complex and uncertain environments, the goal is to optimize the control objectives with minimal human intervention.

2.4 What Kind of AI Technology ...

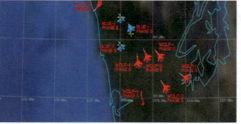

Fig. 2.4 Simulated combat between humans and an intelligent control system [4, 5]

Intelligent control systems have already been used in industry and unmanned machines. The concept of "intelligent agents" has been used in piloting unmanned aerial vehicles and control assistant platforms. In 2016, my colleague Kelly Cohen developed an air control simulation system based on Genetic Fuzzy based AI [4] which competed and won against senior pilot training officers in simulated battle (Fig. 2.4).

With the constant development of cybernetics technology, we have arrived at a new concept: Cyber-Physical Systems (CPS). This has attracted a lot of attention since it has been listed as a key strategy by the United States[3] and Germany.[4] The foundations of CPS in the physical world are the Internet of things (IoT), ubiquitous computing, and execution mechanisms. These define the functionality of systems and are the basis of perception and feedback. Resources, relationships, and references form the basis of operating physical systems and should be managed in cyberspace. Communication, computation, and control in CPS are the technical means to manage the physical world, and comparison, correlation, and consequence are the core means of analysis. The ultimate goal of CPS is to manage visualizability, variation, and value in cyberspace (Fig. 2.5).

CPS can be summarized as collecting, storing, modeling, analyzing, mining evaluating, predicting, optimizing, and collaborating on large amounts of data of an entity space object, environment, or activity, and combining this with design and testing to produce deep integration and real-time interaction with entity space. Cyber space (including the combination of individual space, environment space, group space, activity space and deduction space stated before) guides the specific activities of physical space through the comprehensive utilization of cyberspace knowledge, realizes the accumulation, organization, growth and application of knowledge, and then through self-perception promotes smart systems.

---

[3] In August 2007, The United States' President's Council of Advisors on Science and Technology (PCAST) published a report titled "Leadership under challenge: Information technology R&D in a competitive world" [6]. This made "Network Information Technologies and Physical World Connections" one of eight core technologies.

[4] In October 2012, German Industry 4.0 working group formally proposed to the German government the implementation proposal of Industry 4.0 and made Cyber-Physical Production Systems (CPPS) the key technology of intelligent manufacturing systems [7].

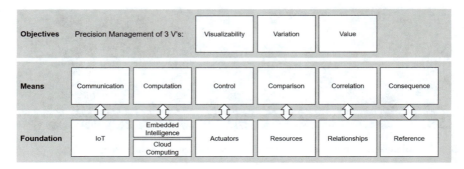

Fig. 2.5 Design guideline for CPS

The essential purpose of a cyber-entity system is to construct a mapping from physical space to cyber space (Fig. 2.6). Physical space consists of the elements and activities that create the physical world, including environment, equipment, systems, communities, and human activity. Cyber space is the synchronization and modeling of physical space, reflecting it by data-driven mirror models, simulating relationships through the establishment of group space, environment space, activity space, and deduction space and how they change over time. We can simulate, evaluate, deduce, and predict activity in the physical space, form decision-making knowledge, and create a complete knowledge discovery system.

CPS is a multidimensional intelligent technology system, based on big data, networking, and high-performance computing. With smart perception, analysis,

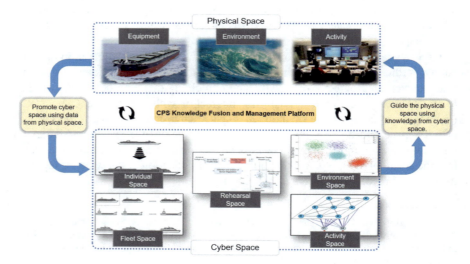

Fig. 2.6 CPS diagram of the relationship between physical space and cyber space in a CPS

data mining, evaluation, prediction, collaboration, and optimization at is core, CPS integrates computing, communication, and control to achieve in-depth collaboration.

The core of CPS lies in data-driven models, analyzing the differences between objectives and the results of activities in the environment, adopting collaborative solutions, and promoting the development of intelligentization. Compared to bionics and statistics, cybernetic experts pay more attention to information-action closed loops, focusing more on information processing and convergence, that is how to maintain the stability and accuracy of decision-making within the domain and protecting the system from situations outside the boundary.

2.4.4 Industrial AI Isn't Just Algorithms, But the Integration of People, Things, and Systems

We have just discussed which AI technology is suitable for industry such as neural networks, but I think that an Industrial AI system need to connect people, things, and systems together. Before the maturity of information technology, people had to manage things and systems. The information systems were mainly apprenticeship, corporate culture, and institutional discipline. The rise of IoT technology connects people and things, using things to prompt people when there are things to do. But this concept is not yet complete, so IoT applications have not brought obvious changes to the industry, mostly because IoT technology only connects people and things rather than changing or optimizing systems. Industrial AI needs to connect people and things through systems, change from experience-based operation to evidence-based operation, and constantly optimize the system in the process. In this process, system and person are connected through mobilization and enterprise applications which break geographic boundaries and make collaboration more efficient. The connection between systems and things depends on CPS, and technical tools such as data technology, analytic technology, platform technology, and operations technology.

The requirements of improving an industrial system can be thought of as the three W's: the first is Waste Reduction which is reached through training and process improvement such as Total Productive Maintenance (TPM), Lean, and Six Sigma. The second is Work Reduction which is accomplished through automation and integrating manufacturing systems. The third is Worry Reduction, which is the process of managing uncertainties we cannot see and problems we might not understand. This is accomplished by AI. We can continuously explore the relationships between people, systems, and things by establishing and managing data resources, carrying out multidimensional reference analysis, and use all this to finally increase the resilience of regulation and reconfiguration.

For example, when we find the correlation between equipment failure and operational parameters, we can optimize equipment usage and maintenance, thus achieving zero-downtime equipment operation. When we find the correlation between operational parameters and final product quality, we can monitor and adjust parameters

Fig. 2.7 Industrial Intelligence system integrating people, things, and systems

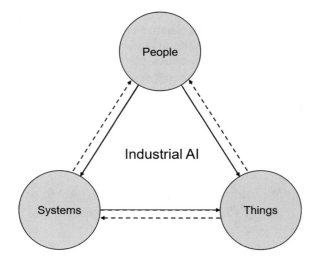

to optimize for quality and achieve zero-defect product production. The same can be done for optimizing minimal energy consumption or minimal material usage to achieve zero-waste production.

2.5 When Machine Intelligence Meets Industry

On March 1st, 2018, The US Center for Strategic and International Studies (CSIS) released the report titled "A National Machine Intelligence Strategy for the United States" [8]. With the rapid development of Machine Intelligence (MI) in various fields over the past five years, the United States is now entering a critical period of technological change. The report aims to promote the safe and reliable development of MI technology and gives recommendations on how the United States can maintain a competitive advantage in the field. Specific measures are put forward for the formulation of MI in the scope of national strategy: R&D, personnel training, data environment, legal policy, risk management, and strategic cooperation.

The report also sets two general goals for the future development of MI. One is to promote the safe and reliable development of MI technology. Specific measures include providing financial support for MI research and development (R&D), training a new labor force in the MI era, and establishing an open and flexible data ecosystem to promote the development of MI. The second goal is to maintain the global leadership position of the United States in MI by formulating reasonable policies to promote MI innovation and applications and establishing cooperative relationships with other countries.

In order to define MI's relationship with humans, the report compares human intelligence with machine intelligence in a way that coincides with the views I have

2.5 When Machine Intelligence Meets Industry

put forward in many speeches and past books. The source of human intelligence comes from observation, learning, and culture, and would be considered experience-based. MI comes from data and algorithms relative to data, paying more attention to objective laws and facts within data. Human intelligence is more adapting to environments with high uncertainty and low predictability, and is more competent in judgement, creativity, and understanding. MI is more adaptable to environments with abundant data, more dimensions of variables, and more predictability, able to perform deterministic tasks accurately and effectively. But we must realize that MI is limited to solving pre-defined problems. Although there are breakthroughs in self-learning algorithms (such as reinforcement learning and transfer learning), the learning process of machine translation is still based on defined loss functions.

Regarding strategies of applying MI, the report highlights the potential of MI tech in training new kinds of labor forces. In another book of mine, "Cyber-Physical System: The New Generation of Industrial Intelligence" [9], I brought up some core contradictions to be solved in the era of Industrial Intelligence. On one hand, we need to increase the efficiency of productivity by aiding people in completing dangerous and repetitive work, and also to enhance the efficiency of knowledge's access, utilization, and transmission by people. Because each country has a different manufacturing culture, the application and emphasis of machine learning will be different by country. But no matter what, it will help solve the contradiction between people to innovate and share information sustainably.

People's renewed attention to the field of AI stems from recent breakthroughs in deep learning. Behind this upsurge of AI is not a revolutionary breakthrough but a limited improvement of existing methods that does not go beyond the scope of neural networks. There is a suspicion of over-hype. The development of the AI industry faces the risk of a bubble, which is mainly reflected in the large quantity of investment supply opposed to the small quantity of project supply, the high expectations of the market for venture capital, the influx of large amounts of funds into hot areas, and the fact that the actual product experience is not that good.

In 2013, GE put forward the concept of "Machine + Minds." It is the idea of combining smart devices based on machine-based algorithms and embedded devices in a large data environment. It creatively distinguishes the difference between knowledge creation and knowledge utilization. In March 2018, Accenture and Horses for Sources (HfS) Research released a report titled "The Future Belongs to Intelligent Operations" [10] which surveyed 460 executives worldwide including 30 Chinese companies. All in all, they believed that enterprises needed to focus on five elements: innovative talent, data support, applied intelligence, cloud enablement, and an intelligent ecosystem. Based on data insights, they want to enhance business results and improve customer experience as to cope with the tremendous impact of digital change.

In the past five years, we've seen many applications of AI in the industrial domain, including machine vision detection, virtual metrology, predictive diagnosis, and human-computer collaborative assembly. We can all feel there is a coming era of Industrial Intelligence, but should also realize the shortcomings of AI in industry. I hope this book can share some of our experience accumulated in the practice of

Industrial Internet for more than ten years, clarify the differences and relationships between AI and MI, introduce the systematic methodology and development process of Industrial Intelligence, and help enterprises better understand how to apply Industrial Intelligence to create value transformation.

On February 11th, 2019, President Trump signed the American AI Initiative, an important national strategy for AI development in the United States. It includes investment in AI, opening up government data resources, building relevant standards, using AI to respond to employment crises, and formulating international policies. These five aspects of the initiative set out the direction of AI development in the United States for the future. It also expressed the exclusion of transnational acquisitions of key AI technologies from potentially threatening countries.

In this competitive and militant executive order, five principles and six strategic objectives are put forward to ensure that the United States remains the world leader in the field of AI. It is hoped that the government's regulatory system, promotion of technical standards, encouragement of innovation, and training will help accelerate the development and application of AI technologies. However, critics pointed out that although the policy set many goals and ambitions, it was ambiguous about specific details and never mentioned where the funds would be spent. While mainly focusing on smart cities and medical fields, it does not focus on industrial applications.

At the same time, Chinese AI industry has also been rising in recent years. Internet giants represented by the acronym BAT (Baidu, Alibaba, Tencent) have invested heavily in AI research and development, and AI unicorn startups have begun emerging. As a result, the United States feels competitive pressure from China's rapid development in AI. In recent years, China's startups in AI mainly focus on speech recognition, image recognition, video recognition, and social interaction, such as TikTok. They are widely involved in the financial industry and shared economy and other areas of life services with large markets and high demand. While there is a lot of progress made in core AI technology, frameworks, and prototyping, the development of architecture, system support, network infrastructure, datasets, and standardized technology in China is behind the United States.

2.6 Differences Between Industrial AI and AI

The differences between Industrial AI and AI (Table 2.1) in general are not only reflected in the application field, but also in functional requirements and algorithms. But first, we need to understand the definitions of the two.

Artificial Intelligence (AI) is a cognitive science with rich research in the areas of imaging analysis & machine vision, natural language processing robotics, and machine learning, etc. AI has been perceived as a black art and often lacks of compelling evidence to convince industry that these techniques will work repeatedly and consistently with a sound return on investment.

Industrial Artificial Intelligence (Industrial AI), is a systematic discipline which focuses on developing, validating and deploying various machine learning algorithms systemically and rapidly for industrial applications with sustainable performance. Intelligent systems in the industrial domain, and is a method to train systems. It is systematic, speedy, and sustainable. Because of its convergent and efficiency-driven features, energy efficiency and safety in industrial production and equipment can be improved along with transportation safety and machinery stability. The application focuses on industrial equipment, manufacturing equipment, transportation, energy industry, production equipment, and automation (Table 2.1).

AI is beginning to be applied to industrial systems on a large scale, but in order to realize smart systems, the following five requirements must be present:

Systematic: The systemization of technology and application levels requires the establishment of clear and consistent protocols to classify the task boundaries and interfaces of Industrial Intelligence at different levels. From the component level, equipment level, system level, and even the community level, there must be a systematic approach to leveraging the entire system to achieve value enhancement.

Table 2.1 Differences between AI and Industrial AI

	AI	Industrial AI
Definition	A trial-and-error judgement-driven technology applied to NLP, image processing, automatic reasoning, robotics, and so on. It can be applied to a wide range of fields such as medical treatment and business, but does not have a successful usage case in engineering	Systematic training and methods that realize AI applications in the industrial domain are speedy, systematic, and sustainable. Different people would use the tools to get the same results, and AI standardization could be advanced
Function	Divergent and opportunity-driven situations such as autonomous driving, economy sharing, and facial recognition	Convergent and performance/efficiency-driven situations based on improving from an original basis, such as improving production efficiency, improving quality, reducing energy consumption costs, improving equipment stability, and improving automobile safety
Application Areas	Social networks, financial sector, medical industry, among others	A broad range of industrial applications including industrial equipment and manufacturing, power grids, power generation equipment, transportation and logistics, medical systems, etc.
Algorithms	Machine learning, deep learning	Deep learning, broad learning, fuzzy learning, augmented learning

Standards: It's not just protocols that need to be standardized, but also the methodology, process, measurement, modeling processes, data quality, model evaluation, fault tolerant mechanisms, prediction-based operating procedures, and uncertainty management among other things. It's especially important to standardize the decision-making process. If something is not able to be integrated with existing standards, it will be hard to integrate technology into industry and produce value.

Streamline: The workflow of the development and implementation of Industrial Intelligence systems is based on systematic methodology, the process of the intelligent industrial system inputting information and outputting decision-making, and the process of the industrial system interconnecting to realize the rapid deployment of intelligent applications.

Speed: Although problems in industrial systems are clear, there are many systems to quickly build, model, validate, and deploy to solve fragmentation problems and to meet customized requirements of rapid response.

Sustainable: Similar to the need for interpretability and certainty of AI predictions, sustainability refers to achieving the same model and results from the same set of data. Without this interpretability, it will be difficult to manage standardization and consistency within manufacturing systems.

The objective of standardization, systemization, and streamlining is to realize speed and sustainability. Different countries have different manufacturing philosophies to achieve this. The Japanese take the knowledge learned from the problem-solving process and pass it down through training to avoid and solve problems. Germans solidify the experience of solving problems into equipment, and use it as the carrier of knowledge to solve and avoid problems automatically. Americans understand problems through data, acquire knowledge, then finally solve and redefine problems, thus a closed-loop system.

Figure 2.8 illustrates the difference between Industrial AI, machine learning, expert systems, and human experience. Expert experience refers to experience acquired by technicians after long-term practice. For example, in the fault diagnosis of rotating machinery equipment, experienced experts can accurately judge and locate the fault only by their human ears. However, this kind of experience has great challenges when it comes to knowledge transfer. Some of the experience disappears with the departure of personnel, which leads to unsustainable guarantees of enterprises solving problems. Additionally, this experience has its limitations on invisible problems. For example, no experienced expert can accurately assess the current health status of the equipment.

Expert systems are systems based on localized professional knowledge and centered on knowledge base and inference engines, so it is difficult to deal with the uncertainty caused by the change of environment and working conditions. Performance improvement requires periodic updates of the system. In general, the systems developed by AI and machine learning have greatly improved the accuracy of solving problems compared with the first two methods, and have fantastic learning ability. However, the robustness of the system is insufficient in the face of changeable

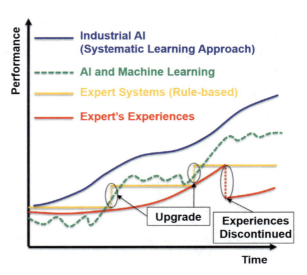

Fig. 2.8 Comparisons of expert expertise, expert systems, AI, and Industrial AI [11]

working conditions and diversified data. Future Industrial AI is based on multidimensional learning, and has the characteristics of transfer learning systematically, so the performance of the system will be steadily improved.

2.7 Challenges of AI in Industry

Although AI technology has made breakthroughs in many applications, there is still a big gap between large-scale usage in industrial scenarios. This is because industry and manufacturing value stability, standardization, accuracy, and repeatability, as well as mechanization, processes, operations, and close integration of process requirements. Before AI technologies can be fully integrated into industrial systems, it is necessary to overcome the challenges of reproducibility, reliability, and security.

2.7.1 Reproducibility

On February 15th, 2018, Matthew Hutson pointed out in "Science" [12] that most of the algorithms published in the papers, especially machine learning algorithms, had not been verified as reproducible, so machine learning algorithms could only be regarded as theories and hypotheses, but not systematic. Matthew holds that there are three weaknesses in using reasoning as a logical tool: (1) relevance cannot prove causality, (2) uncertainty is inevitable and, (3) the probability of errors due to knowledge limitations.

One of the factors leading to non-reproducibility is that it directly replaces the probability of strong inductive reasoning with deterministic results, completely ignoring the most important feature of strong logical reasoning. There are limitations and deviations in the process, and the deviations will directly affect the reliability and safety of industrial systems.

2.7.2 Data Issues

With regard to data, AI technology faces five major challenges and limitations: (1) training data is heavily dependent on manual work, otherwise it is difficult to obtain a large and comprehensive training data set, and the quality of labeling is heavily dependent on human experience and ability. (2) The transparency of the model needs to be improved since AI algorithms cannot explain how conclusions are reached step-by-step. (3) Models are not very general, and it is hard to replicate from one application to the next. This means lots of money and energy is needed to train new models for new problems. (4) The risk of deviation in data and algorithms, much like the differences between societies and culture, requires extensive steps to solve, (5) It is difficult to reach agreement on data privacy and attribution.

Considering the importance of data, it is important for enterprises and organizations to create data strategies, have services run in datacenters, and attract data experts.

In technology development, enterprises and organizations must develop sound data maintenance and management to achieve modern software development specifications. The most difficult problem is the "last mile" of customer service which ensures that AI can be implemented into business processes, products, and enterprise services.

We may have had the wrong impression that we acquired enough analyzable data, but in our experience in cooperating with many enterprises, this is not the case. We might have a lot of data, but we still have problems in irregular data acquisition, lack of key parameters, misalignment of variable timing, lack of labels such as working conditions and maintenance records, and low data quality in almost every enterprise. To solve these problems in industrial data, it is necessary to understand and manage the challenges posed by the "three B's" outlined below.

Bad Quality: In industrial big data, data quality has constantly been an issue for many enterprises. This is mainly limited by the means of data acquisition in the industrial environment, including sensors, communication protocols, configuration software, and other technical constraints. The management technology of data quality is hard work for any enterprise.

Broken: To be useful to industry, data can't just be numerous—it must be comprehensive. When using data models to solve a problem, we need to obtain comprehensive parameters related to the object being analyzed, and also obtain key parameters to fragment the analysis process. For example, parameters such as temperature, air density, inlet and outlet air pressure, power, and so on are needed to analyze the

2.7 Challenges of AI in Industry   27

performance of aircraft engines. When any of them are missing, a complete performance evaluation and prediction model cannot be established. Therefore, before data collection for enterprises, there needs to be a clear plan of the object and purpose of analysis in order to ensure the comprehensiveness of the data obtained. It would be a problem if a lot of money was spent to accumulate a large amount of data, only to find that the problems cannot be solved.

Background: Besides analyzing the surface statistical features reflected by data, we should also pay attention to the hidden correlation in the data. When analyzing the correlation hidden below the surface, we need reference data to compare. This is called "labelling" in data science. This kind of data requires work condition settings, maintenance records, task information, and so on. The amount of auxiliary data is small, but it plays a vital role in data analysis.

2.7.3 Reliability

According to the different requirements of reliability in different domains and applications, AI can be roughly divided into mission-critical and non-mission-critical applications. Currently, most AI products on the market do not require strict system reliability. As long as a certain threshold of usability is reached, the occasional errors and problems can be tolerated without serious consequences in non-mission-critical applications.

After the release of the iPhone X, its Face ID unlocking feature—a major selling point—drew criticism from users saying that it could not be used under certain circumstances. But even if the unlock fails, it can be restored to normal through a system reset, which does not cause harm to users. Recommendation systems use AI to judge user preferences and recommend content that may be interesting based on the user's behavior and user data. Shared bicycles in China may not unlock when a user scans the Quick Response (QR) code, but there are frequently other bikes around. These systems do not require high accuracy and reliability.

For critical applications, if a system has even a small chance of failure, it could lead to serious consequences, causing property loss or even harm to human beings or social stability. An example of this is the intelligent driving industry: it is expected that this industry will globally reach $9.5 billion by 2020. The industry is facing serious challenges in guaranteeing safety. The first fatal drone crash occurred in March of 2018, and an Uber autopilot test car hit and killed a woman in May of 2018. In the latter case, the autonomous driving system did not take any braking measures before the accident occurred, and the safety guard did not take over the driving in time.

Although more details about these accidents have not been revealed, two possible reasons could be deduced from the structure of the autopilot system: (1) the sensor and detection equipment cannot perceive pedestrians in time, or (2) the decision-making mechanisms did not respond urgently enough to a possible accident. This accident shows us that the current unmanned driving technology for uncontrollable emergencies is insufficient. Currently, driving casualties are recorded at a person's death for every 15–20 million miles driving, which is below the level of someone dying for every 86 million miles a person would drive. In terms of system stability, Waymo, an unmanned driving project not yet released by Google, has reached an average of 5600 miles before requiring manual intervention, while Uber requires frequent manual intervention.

Whether it is true autonomous driving technology or a driving assistance system, its high demand for reliability and intolerance of failures makes it difficult for the technology to truly enter the market before exceeding the human driving level. These challenges are the same in industrial systems. If we are to use AI to manage the operation of an entire system, we must be very cautious about mission-critical work, which requires not only breakthroughs in the accuracy of models and algorithms, but also security boundaries and management of uncertainties in system design.

2.7.4 Safety/Security

As shown in the opportunity space of Industrial AI, people are always passionate about being able to perform tasks with visible results, and often ignore the invisible but more important issues such as managing safety and risk. In March 2018, a fatal car accident occurred on the Bay Area Expressway: a Tesla Model X driver ran out of control, crashed into the median, broke into two pieces, and then exploded and the lithium batteries caught fire. The car owner was declared dead, and this disaster took six hours to clean up. This accident not only exposed the safety problems of new energy vehicles, but caused worries about the safety of batteries and their impact on the environment and social order. Automobile manufacturers guarantee both invisible safety, and visible high-tech brings drivers unprecedented experience.

In mid-March 2018, the New York Times, the Guardian, and Observer revealed that a company called Cambridge Analytica obtained the personal records of 50 million Facebook users through third-party applications, and subsequently conducted large-scale data analysis on the users to psychological profiles for them [13]. They used this analysis to understand their preferences and political tendencies, and influenced their political tendencies and final votes by controlling the information they saw, effecting the US electoral process. This incident demonstrated how big data can be used for political purposes.

2.8 New Opportunity Spaces for Industrial AI to Realize Industrial Value Transformation

In *CPS: A New Generation of Industrial Intelligence,* we compared the first three industrial revolutions with the current industrial revolution that is taking place. This

2.8 New Opportunity Spaces for Industrial AI ...

industrial revolution is known as the Smart Revolution. The bottleneck of productivity that needs to be solved lies in the invisible part of technological factors, since the efficiency of knowledge generation and utilization can no longer meet the requirements of production systems. Production systems driven by human knowledge and experience reached the boundary of productivity, where it is difficult operate and communicate with the maximum efficiency. Restricted by human experience and knowledge, a large part of production system's value driven by people's decision-making has not been utilized. This is mainly reflected in the speed of knowledge acquisition, the depth of ability, and the scale of application. The fourth technology revolution must bring the following core functionalities to industry:

(1) Promote knowledge to become the core factors of production in production capability, and make the discovery, usage, and transfer of knowledge more efficient and wider in scope, allowing for improvement.
(2) Re-optimize the value chain of an organization, so all links in the industrial chain are optimized for value delivery and organizing production activities in an efficient and collaborative way.

Understanding this is helpful for us to recognize emerging technology concepts. The essence of the birth of new technologies such as AI, big data, IoT, Industrial Internet, and cloud manufacturing is to improve the efficiency of knowledge productivity factors and value delivery.

We can separate problems into visible problems and invisible problems to analyze the current application domains of Industrial AI. Currently, the development of AI technology focuses mainly on solving problems in the visible world, where the technology and application is to replace human beings to complete repetitive jobs that humans don't want to do. Boston Dynamics has special robots that are able to do dangerous tasks such as military or rescue operations instead of humans. In recent years, medical imaging technology has been continuously progressing. A lot of medical imaging data such as X-ray, ultrasound, magnetic resonance imaging (MRI), and computerized tomography (CT) scan data has been accumulated. These can be used as data to train learning models to assist doctors to diagnose and treat, reducing manual misjudgement.

AI avoids the application of visible problem areas and needs to accomplish tasks that people do not do well enough. Translation systems based on deep neural networks such as Google Translate and Baidu Translate, and semantic understanding and dialog systems such as Siri and Alexa have achieved success, marking a breakthrough in the theoretical method of effective communication between humans and computers using natural language. In the future, with the development of contextual representation and logical reasoning, knowledge maps will continue to expand. Humans will achieve human-computer dialogue through natural language, better communication, and achieve the goal of humans avoiding tedious interaction.

Unmanned and autonomous driving technology detects the surrounding environment through its sensor system as it drives, including the movement of pedestrians and other vehicles, road and traffic signs, real-time information and road condition information, and can plan optimal routes to the destination, detecting and avoiding delays. Currently, Google and Uber are developing this kind of technology.

AI needs to help people understand the problems and opportunities that do not exist in solving invisible problems. With explosive growth in data, it is a great challenge to help users understand the most valuable information in a short amount of time. Intelligent search and using large data environments provide more intelligent, customized, and human-friendly information retrieval services through role registration, preference recognition, and semantic understanding. Smart search combines speech recognition, image recognition, positioning functions, and so on. It's already widely used in many forms by companies such as China's Taobao, Baidu search, listening to songs, mapping software, weather software, and other products.

Another task in this area is how to use big data environments to assess potential risks. In the financial big data environment, the establishment of intelligent credit information plays an important role in assessing risk; in the medical field patient's records are used to assist diagnosis and treatment.

For problems encountered in manufacturing systems, I often classify them as visible and invisible. The ways to deal with these problems can also be divided into solving problems after they occur, and avoiding problems before they occur. The invisible problems in production systems include the deterioration of equipment performance, the loss of accuracy, wear and tear of parts, and resource waste. Visible problems are often caused by accumulations of invisible factors, such as the decline of equipment leading to shut down, the lack of accuracy leading to product quality deviation, and so on. Like an iceberg, the visible problem is only part of the entire issue, while the hidden problems are larger but unseen. The function of big data and AI technology is to predict the invisible problems in production in order to obtain the means to avoid and solve the invisible problems. This will create carefree manufacturing.

Following this idea, we can further divide the opportunity space of Industrial AI into four parts (Fig. 2.9).

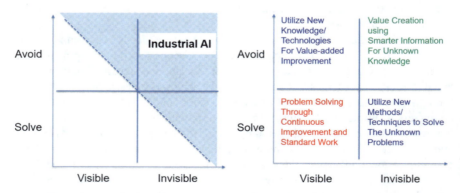

Fig. 2.9 The four quadrants of Industrial AI opportunity space (left); industrial big data in industry (right)

The first opportunity space comes from solving visible problems in the production system. There are still lessons that manufacturers need to understand in this space, such as long product delivery cycles, excessive pollution, and wastes of resources. In order to solve this opportunity space, we need continuous lean improvement of the production system and continuous standardization of the system.

The second space focuses on avoiding visible problems, using data to mine new knowledge, finding the causes of deeper problems, and improving the original production system products. For example, a factory can use 6-sigma diagnostic tools to analyze the causes of equipment failures throughout the year, find the most frequent failures and influential factors, and establish corresponding management mechanisms to intervene.

The third space lies in the dominance of invisible problems. It's important to use innovative methods and technologies to find solutions to these invisible problems and to enhance the theoretical limits of current manufacturing systems. What is needed is more in-depth evidence, relevance and causality mining, and through the establishment of relationships the ability to quantify the original invisible processes. For example, the health deterioration of equipment can be solved with predictive maintenance of equipment through the analysis of equipment mechanisms, operation environments, and operation parameters.

The fourth space is finding and satisfying the invisible value gap and avoiding the influence of invisible factors. This part needs to use large data analysis to expand the scale of relationship building to create a higher dimension of knowledge, and through it achieve a closed-loop integration of the manufacturing system industry chain, and optimize the upstream and design side of the system to avoid invisible problems.

Currently, the vast majority of manufacturing enterprises are concerned about the improvement opportunities in the first and second spaces. However, Industrial Intelligence technology should be applied to solve and avoid invisible problems— opportunities in the third and fourth space. Only then can we achieve breakthroughs in production efficiency and product competitiveness of manufacturing systems and realize the transformation of manufacturing systems to intelligent manufacturing systems. This book will focus on the invisible world: the areas of industrial AI and applications that lie in the shadows. This especially concerns using AI for human prediction and avoidance of problems that haven't even occurred yet.

References

1. Lee J (2015) Industrial big data: the revolutionary transformation and value creation in INDUSTRY 4.0 Era. China Machine Press
2. Sharma J, Angelucci A, Sur M (2000) Induction of visual orientation modules in auditory cortex. Nature 404(6780):841

3. McKinsey Global Institute (2018) Notes from the AI frontier: insights from hundreds of use cases. Available from: https://www.mckinsey.com/~/media/mckinsey/featured%20insights/artificial%20intelligence/notes%20from%20the%20ai%20frontier%20applications%20and%20value%20of%20deep%20learning/notes-from-the-ai-frontier-insights-from-hundreds-of-use-cases-discussion-paper.ashx
4. Ernest N, Carroll D, Schumacher C, Clark M, Cohen K, Lee G (2016) Genetic fuzzy based artificial intelligence for unmanned combat aerial vehicle control in simulated air combat missions. J Def Manag 6(1):2167–0374
5. Lisa Ventre (2016) Beyond video games: new artificial intelligence beats tactical experts in combat simulation. UC creative services. Available from: https://magazine.uc.edu/editors_picks/recent_features/alpha.html
6. Marburger JH, Kvamme EF, Scalise G, Reed DA (2007) Leadership under challenge: information technology R&D in a competitive world. Executive Office of the President Washington Dc President's Council of Advisors on Science and Technology
7. Digital Transformation Monitor (2017) Industrie 4.0, Germany
8. Carter WA, Kinnucan E, Elliot J, Crumpler W, Lloyd K (2018) A national machine intelligence strategy for the United States. Center for Strategic & International Studies
9. Lee J, Qiu B, Liu Z, Wei M (2017) Cyber-physical system: the new generation of industrial intelligence. Shanghai Jiao Tong University Press
10. HfS Research, Accenture (2017) The future belongs to intelligent operations. Available: https://www.accenture.com/_acnmedia/pdf-70/accenture-intelligent-operations-research-web.pdf
11. Lee J, Davari H, Singh J, Pandhare V (2018) Industrial artificial intelligence for industry 4.0-based manufacturing systems. Manuf Lett 18:20–23
12. Hutson M (2018) Artificial intelligence faces reproducibility crisis. Science
13. Kevin Granville (2018) Facebook and Cambridge analytica: what you need to know as fallout widens. The New York Times. Available: https://www.nytimes.com/2018/03/19/technology/facebook-cambridge-analytica-explained.html

Chapter 3
Definition and Meaning of Industrial AI

3.1 The Beginnings of Industrial AI

The concept and term of Industrial AI was first introduced by Prof. Jay Lee of NSF I/UCRC on Intelligent Maintenance Systems (IMS) at the University of Cincinnati. As industrial technology has developed, it has focused on increasing efficiency, reliability, reducing costs, and ensuring quality. More emphasis is placed on predictive maintenance and productivity management.

Although some specific forms of AI are applied and solved in the development of specific industrial problems, the field of Industrial Intelligence has not been formalized. Research in this "narrow intelligence" is still in its infancy (Fig. 3.1).

In the 1950s, Alan Turing put forward the Turing test, and the definition of AI formally ascended to the stage of history at the Dartmouth Conference, but at that time there was no knowledge of "intelligence" in the software industry. During this time in the reconstruction period after World War II, industry developed rapidly, and many wartime industrial management systems were introduced to civilians—Standard Process Control (SPC) theory was one such system.

SPC was proposed by Dr. Walter Shewhart in the 1930s as a technology which could manage production processes by statistical analysis. SPC technology could warn of abnormal trends in key parameters during the production process, as to improve product yield. During World War II, the United States established quality management standards based on this technology, which played an important role in ensuring the quality and timely delivery of military products. Because of the strength of the American economy and its manufacturing capacity, SPC did not play a large role in American civil industry. Instead, it became widely promoted and valued in Japan, and was an important factor in their remarkable post-war rise in product quality and production efficiency. Successful application of SPC technology did not only increase efficiency, but also introduced the idea of using data-driven methods to manage uncertainty in manufacturing.

In the 1980s, the Six Sigma management system was introduced by Motorola and promoted by Jack Welch, then the legendary head of General Electric. It quickly

© Shanghai Jiao Tong University Press 2020
J. Lee, *Industrial AI*,
https://doi.org/10.1007/978-981-15-2144-7_3

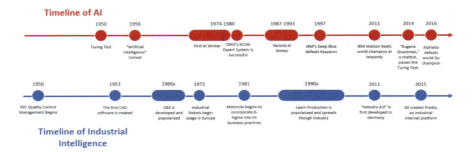

Fig. 3.1 The development of AI and Industrial Intelligence

became a classic management method that still has influence today. Six Sigma is based on statistical theory and integrates a series of management tools to control product quality and reduce change in the production process. Six Sigma is a set of management systems which are mature, and there is even an entire industry that has been formed around it. In recent years, Six Sigma and Lean manufacturing have been combined to create the "Lean Six Sigma" management method. This method combines production process improvement, resource effectiveness management, and quality control. Whether it is Six Sigma or Lean Six Sigma, they are management systems, not intelligent systems. They focus on managing organizational and process uncertainty, the core of which is still people.

Parallel to the development of quality control theory is automated manufacturing technology. From the 1940s to the 1950s, Computer Numerical Control (CNC) machine tools were invented and commercialized, which played a large role in the manufacturing industry and greatly improved production efficiency and accuracy. The development of microprocessors in the 1970s further reduced the cost of CNC machine tools, increasing their usage in the industry. Beyond CNC machine tools, industrial robots were developed during the same period. In 1956, American inventor George Devol and physicist Joseph Engelberger invented the world's first industrial robot, Unimate.

The technology of numerical control machine tools and the gradual popularization of integrated circuits increased the pace of the design and manufacturing of faster and more concise industrial robots. The manpower of industrial manufacturing systems is gradually being reduced, and the degree of automation is steadily increasing. However, the field of automation and control in industry has always been self-contained. The principal problem is that of functionality: how the machine replaces human beings to complete an action or series of actions without caring about whether the machine itself can think. Today, control theory tends to combine with AI. For example, autonomous driving technology is an automated product that integrates AI.

The precondition for putting forward the concept of Industrial AI lies in the conditions of engineering. If we know the history of AI, we will find that the subject can

3.1 The Beginnings of Industrial AI

be separated into two fields: General AI and Narrow AI. General AI emphasizes that machine capability can surpass the development of human intelligence. Intelligence is the focus of technology in this field, and infrastructure such as servers, sensors, and software durability is a prerequisite for AI technology to be effective. With the increasing maturity of industrial equipment automation, the deployment of sensors, and the improvement in computing power, the concepts of industrial Internet, industrial IoT, and Industry 4.0 are emerging endlessly. The time is ripe—no matter what the future industry will be called, people are ready to embrace change.

However, the are many challenges in applying general AI directly to the industrial domain: machine learning models cannot train themselves with the experience and knowledge of domain experts. Throughout its history, Industrial Intelligence was only a technical means for enterprises to make industrial systems intelligent. Even today, improving efficiency, reducing costs, and guaranteeing quality remain unchanged themes. Therefore, a closer approach to the application of AI is to apply narrow AI technology to solve problems with discrete definitions and definite boundaries in the industrial field. AI scientists should combine domain knowledge with data generated by equipment to train Industrial Intelligence applications with specific functions and realize intelligence in specific problem areas.

From the two timelines of AI technology development and Industrial Intelligence development, we can see that their development is similar and overlapping; however, there is little overlap in application. For example, the application of AI in manufacturing quality has not been widely implemented until now, but the problem of quality was put forward in 1924. Although Lean manufacturing was widely practiced in factories in the 1990s, data-driven intelligent Lean manufacturing rarely replaces hands-on Lean manufacturing. Thus, we can see that the application of AI technology in industry lags far behind the development of the AI technology.

What changes can AI bring when applied to the development of Industrial Intelligence? Quality problems used to be handled by SPC, which was based on effective measurements and would identify and solve visible problems quickly. Can Statistical Process Prediction (SPP) be achieved after the application of AI technology so that invisible problems and process anomalies can be measured and detected in time? If SPP can be achieved, many problems in quality will be solved and avoided. In terms of equipment management, from statistical reliability-centered maintenance (RCM) to predictive maintenance (PdM), zero downtime (ZDT) can be achieved. So, we say that Industrial Intelligence is taking place, not because of the number of new and disruptive technologies, but because we are finding suitable scenarios and objects to engineer and systematically integrate existing technologies. For example, at the International Manufacturing Technology Show (IMTS) in Chicago in 2018, Fanuc introduced a ZDT machine tool control system which integrates IoT and predictive analytics [1] on the basis of the concept proposed by Prof. Jay Lee of NSF I/UCRC on Intelligent Maintenance Systems (IMS) in 2001.

In Fig. 3.2, we summarize the five stages in which industry is moving towards more intelligence. We believe that these five stages are the process that must be experienced, and it is very difficult to achieve success across each stage.

Level 1: 5S and Kaizen Model (Hands-on Level)

Level 2: Lean Manufacturing Systems and Six-Sigma (Data Level)

Level 3: Predictive Analytics Tools (Insight Level)

Level 4: Decision Making and Optimization Tools (Knowledge Level)

Level 5: Cyber-Physical Systems (Autonomous Intelligence Level)

Fig. 3.2 Five stages of moving toward smarter industrial systems

The first stage, the "Hands-on" Level, is the practice of staff members. This was first proposed and promoted in Japan. Its core element is called "Kaizen" in Japanese, which when translated to English means "continuous improvement." This refers to doing a good job of tidying up, cleaning, and day-by-day improvements to standardization and continuity. The main tools are the PDCA cycle (Plan, Do, Check, Action) and the organizational culture practiced by all staff within an organization.

The second stage is the Data Level. The Lean management system proposed by Toyota in the 1980s and 1990s and the Six Sigma management system promoted by GE are part of this stage, emphasizing how to build a data-based management system around measurement and statistical technology.

The third stage, the Insight Level, is about predictive modeling and analysis. This relates to the transformation of industry in the United States from 2000 until now. It is to solve closed-loop problems beginning with the data layer, and extending to the information layer, and then to the decision-making layer.

The fourth stage is the Knowledge Level, focusing on the optimization of autonomous decision-making. There are many forms of knowledge systems. In the past, expert systems, physical models, and statistical models were the main forms, but now with the help of intelligent algorithms and big data, a knowledge system with higher dimensional data and more complex relationships can be built. The aim of these knowledge systems is to predict future situations and unmeasurable variables accurately, and then to achieve more optimal and timely decision-making. At this level, what we need to do is transform data and experience into a system that can support decision-making and, more specifically, to transform experience-based decision-making into fact-based decision-making. Although experience can be passed on, it is difficult to do so in a complete and accurate way over a long period of time. Data is easier to pass on because it is more concrete and logical.

The fifth stage, the Autonomous Intelligence Level, is the highest stage of Industrial Intelligence. In this stage, an industrial system can generate new knowledge independently in a closed loop including: perception → analysis → prediction → decision → execution → feedback, so as to further optimize the knowledge system and decision-making system, and realize the autonomy of knowledge generation, utilization, and transmission. For example, nowadays autonomous driving is growing in popularity. I believe it is not enough to achieve autonomous cars—what more people need is a more carefree driving experience. That is, driving without worrying

3.1 The Beginnings of Industrial AI

about the road ahead, congestion, road surface conditions, police survailance, or the risk of car accidents. If there is a pothole on the road and a car passes by it, approaching drivers need to be warned of the risk ahead. Although the vehicles behind this car have not yet reached this place, they should receive notification. Such a scenario is called "worry-free driving."

The same is true in industrial systems—if the correlation between the equipment itself and the historical data of other similar equipment can be established, we can take the initiative to avoid bad operations and constantly find better operations. This is called the value conversion of data: performance optimization, avoiding risks, and avoiding worry.

The book "Industrial Big Data" discusses methods for solving visible problems such as by recognizing big problems from the beginning of productivity. The problems discussed were big enough at the time for us to in invest in collecting data. This is what I did for the first half of my life. Now, in the second half of my life, I have changed my approach because solving problems is not a means to an end—the purpose is to make problems cease to appear. This is what is meant by finding value from hidden problems (or problems that customers are not aware of). It is of great value to avoid a problem before it arises.

Figure 3.3 is the vision I proposed to the National Science Foundation (NSF) when the IMS Center was established in 2001. We had an idea on how to turn the big problems from the first half of our lives into the great values of the second half of our lives. Only by accumulating sensor data and historical data, and then performing detailed and in-depth analysis, could the essence of industrial big data be brought into play. We need to find out the hidden problems, the problems that have not happened, so we can either solve or avoid them. From there we can create value, which is the second half of industrial big data. Competition in the next economic era will take place in an evidence-based economy, not just in a social network or experience-based economy as recognized today. The application of traditional consumer internet and social media is divergent: their relationship relies on the identity of the consumer, targeted advertising, the location of the next business opportunity, and how to stimulate consumer demand. These are all opportunity-oriented strategies.

Foxconn's production line accumulates massive manufacturing process data, and we can find the relationship between key parameter changes and product quality, accelerating the convergence rate and stability of new product investment (NPI). By detecting the control parameters and spindle load changes in CNC machine tool cutting processes, the degree of tool wear and remaining useful life can be predicted as to avoid machining failure and reduce tool cost. This corresponds to the improvement of the "yolk" capability in the "fried-egg" model shown in Fig 3.4.

GE engines save fuel by remotely collecting data and recommending flight operations, saving billions of dollars every year by reducing yearly aviation fuel usage by 1%. When navigating cargo ships, the relationship between steering, ocean current, and fuel consumption is modeled according to weather, wave, and wind conditions, as well as other variables. Depending on the analysis, the route and speed at sea can be optimized to save up to 6% of fuel consumption costs. I call this the "fried-egg" model of manufacturing value. Using this analogy, which is illustrated in Fig. 3.4 ,

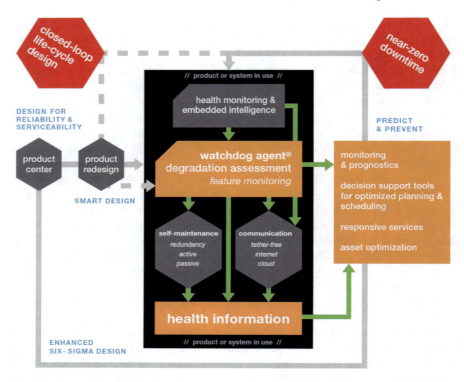

Fig. 3.3 Future Industrial Intelligence system planning, submitted by IMS center to NSF in 2001: from problem solving to problem avoidance

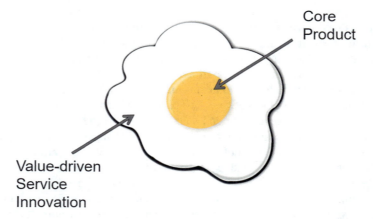

Fig. 3.4 Fried-egg model of industrial value: functional value and service value of products

AI is applied to an industrial big data environment to continuously enhance the core competitiveness and value of the product.

With the continuous development of big data, AI, blockchain, and 5G technology, future generations will benefit from a large number of subdivided economic fields which will serve their needs more accurately. At the same time, in today's internet economy, different kinds of interconnected and overlapping networks such as industrial and group networks will emerge according to people's needs and roles.

We have made great efforts to promote the industrial internet and industrial cloud, but have also encountered many difficulties. Is it more appropriate to establish a proprietary platform focusing on objects and problems? For example, Foxconn Industrial Internet's factory service cloud, robot cloud, and other cloud platforms provide value as service networks that can solve problems and create value for specific industry scenarios.

3.2 The Purpose and Value of Industrial AI

There are differing opinions about the definition of Industrial Intelligence systems in academia and industry. Some have tried to define Industrial Intelligence as a certain technology or solution, but they often ignore several fundamental problems such as "What kind of intelligence do industrial systems need." "What problems and challenges have not yet been solved by previous methods", and "How can AI solve these problems?" There is a Chinese proverb that illustrates this concept: "Every rooster says that the run rises because of its crows." While a rooster might think that it crows and the sun will rise, in reality the sun serves many other functions like stimulating the growth of plants through photosynthesis and providing solar energy. The same is true of Industrial AI. Its purpose should not be for the use of data scientists to prove the omnipotence of AI, to rewrite traditional industrial systems, but should be used to solve those invisible problems within industrial systems that have not yet been fully recognized.

Industrial AI is not just a reuse of general AI technology in the industrial domain. Within industry, the characteristics of fragmentation, individualization and specialization of problems determine that Industrial Intelligence depends on the deep integration of computer science, AI, and domain knowledge. Unlike traditional methods of rule-based or mechanism-based modeling, one of the major advantages of data-driven Industrial Intelligence technology is the ability to establish predictive analysis based on insight and evidence contained in the data, which allows for establishing smart management tools for invisible problems and exploring the relationships between complex things. This way, new knowledge is accumulated to form an intelligent system which can be continuously iterated on.

Generally speaking, we can use four-quadrant graph to categorize problems in manufacturing systems. Problems can be categorized as either visible or invisible (Fig. 3.5). We can wait for problems to happen (or become visible) and then solve them, or avoid them before they occur (while they are still invisible). An example

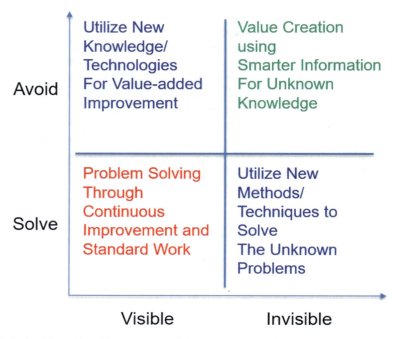

Fig. 3.5 Visible and invisible problems in industrial systems

of this is the difference between waiting for a mechanical cutting blade to break and then replacing it, or using AI to analyze its remaining useful life and replacing it right before it breaks based on vibrations or other data.

Over the past half-century, we have been solving visible problems in industrial systems, accumulating experience and knowledge in the process, and avoiding visible problems through continuous improvement of systems and processes. The span from the first to second quadrant has involved technologies such as testing and measurement, digital manufacturing, and statistical analysis. In the early 1990s, Professor Shien-Ming Wu and his students launched and promoted the "2 mm Project" for the U.S. auto industry. The purpose was to use statistical science to analyze huge amounts of data, and to model the accumulation process of quality errors in order to explain the sources of error and control them. The Six Sigma fluctuation range of the key dimensions in this case had to be less than 2 mm. The main technology used at the time was the application of streams of variation in the multi-stage automotive body assembly process. By modeling the data flow of the success of a complex product flow, the correlation between quality fluctuation and error transmission in the multi-stage manufacturing process was analyzed. Workstations were organized to create different assembly units, and assembly units were further organized to comprise the assembly line for the automotive body production process. Within each assembly unit, each workstation had a dimension variation. The error transfer relationship caused by each assembly component moving to the next workstation.

3.2 The Purpose and Value of Industrial AI

The data generated by a workflow contains three dimensions of correlation: the correlation between quality attributes and different stages of the production line, the interaction between quality attributes in the same production stage, and the relationship between quality attributes and time (generated by the decay of equipment over time). On the basis of these three dimensions, the relationship between key control characteristics (KCC) and key product quality characteristics (KPC) is established and the variation of quality is regulated by controlling KCC.

This method is still widely used in modern manufacturing and is the standard, classical method of quality control. In addition to using data analysis to control process quality, similar analysis methods are also applied to production balance, production scheduling optimization, resource planning, and supply chain optimization of manufacturing systems.

Especially with the rise of digital manufacturing technology, the number and quality of measurable process parameters has increased. Process anomaly detection and electronic Kanban technology allow the anomalies in production to be perceived at the very initial stage. This significantly improves the agility of problem finding and handing. At the same time, the mature application of manufacturing execution systems (MES) enables us to modify manufacturing parameters quickly, improving the speed of decision feedback to execution. Some people equate digital manufacturing systems with intelligent manufacturing, but I think this is inaccurate.

Using traditional measuring, monitoring, and statistical analysis, we can only manage the visual problems that have occurred, but have no way to solve invisible problems. The real value of intelligent manufacturing should be to help us recognize, manage, and avoid these invisible problems.

What are the invisible problems in industrial systems? Taking Fig. 3.6 as an example, each stage in the manufacturing system process is affected by human, equipment, material, process, and environmental problems. These factors (x) can within a design value (μ) produce variations in production (σ) which will result in

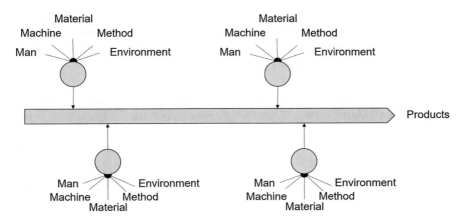

Fig. 3.6 Invisible problems in manufacturing systems

variations of quality (y). These factors collectively would cause variations in the final product quality (Y). In the traditional method, we use Cpk to manage y, and error flow analysis to manage the relationship between y and Y. However, there are still many hidden problems that cannot be managed, mainly reflected in:

- The process factors cannot be measured, and the state of the machine can be unknown. This includes deterioration of equipment performance, deviation of process parameters, and inconsistency of incoming materials.
- The relationship between process factors (x) is not clear, which may cause all of x to be within the control limits while the quality of the process still deviates.
- The relationship between the variations of process factors and the quality of intermediate processes cannot be accurately and quantitatively assessed.

The first purpose of Industrial AI is to make the hidden problems in an industrial system explicit, then to avoid those problems by managing them while they remain hidden. The core technologies include:

1. Measuring process factors that cannot be automatically measured, such as equipment conditions and fault prediction through machine vision, pattern recognition, advanced sensing, and other technologies.
2. Modeling the relationship between process factors and process quality: Multivariate process anomaly detection, virtual metrology, deep learning, neural networks, association rule mining, and other technologies.
3. Dynamic optimization of optimal process parameter settings enables the enhancement of the system resilience and automatically compensates for errors including: optimization algorithms, dynamic error compensation, intelligent control systems, and other technologies.

The second goal of Industrial AI is to accumulate, inherit, and apply knowledge on a large scale.

Human knowledge generation can no longer meet the requirements of the production system. Humans are reaching the peak efficiency of operation and collaboration. Restricted by human experience and knowledge, a large part of the value of production systems has been inhibited by the limitations of human decision making. This is mainly reflected in three bottlenecks: the speed of knowledge acquisition, the maximum ability capacity, and the scale of applications. The management of knowledge through Industrial AI needs to achieve the following two goals to surpass this:

1. Enhance the productivity of knowledge as the core factor of production and make the production, utilization, and transmission of knowledge more efficient at large-scale.
2. Reoptimize the value chain relationship of production organizational factors so that each link in the entire industrial chain can provide services to end-users in an efficient and collaborative manner, focused on providing value.

So, what is knowledge? How should knowledge be modeled and managed? We believe that knowledge within an industrial system is the relationship and operations of objects, environments, and tasks. It is an abstract expression of comparability, relevance, and purpose, and can be summarized by the three R's:

3.2 The Purpose and Value of Industrial AI

- Resources: Knowledge and experience are built on the basis of sample data from observation and results. Data sources can be historical data, sensor data, or human experience data, all of which can combine in a logical way to form a knowledge model. Resources are also the basis of comparability.
- Relationships: Based on the analysis of comparison and correlation, relationships between the visible and the invisible can be found. For example, there are hundreds of sensors generating data in semiconductor fabrication, and historical alarm information and Bayesian networks can be used to establish relationship maps between them. If, perhaps, these hundreds of sensors had strong correlations with a group of five particular sensors—these five sensors could then manage the status represented by the sensor data as a whole. Another example is how, during the operation of engines developed by GE, a relationship between air pressure, air density, combustion temperature, and speed can be established and used to reduce fuel efficiency by optimization.
- Reference: There are two aspects of reference: one is the reference of comparison, the other is the reference of execution. This can also be divided into active reference and passive reference. Reference is also the basis of memory; if a result is used as the point of reference, the purpose is then to find the root cause of its occurrence. If a process is used as the point of reference, then the purpose is to find ways to avoid problems. There is a saying in Chinese: "Using glass as a mirror, you can fix up your clothes. Using history as a mirror, you can learn the cycle of a dynasty. Using people as a mirror, you can understand right from wrong." This ancient proverb summarizes the three dimensions of reference: How sensor data reflects a system (using a normal, glass mirror that reflects reality), how historical data reflects the interaction of data (using history as a mirror to learn patterns), and how people understand and interpret the data (using people as a mirror to understand the real effects of things).

Industrial AI can achieve a significant improvement in the management of the above three R's. It can acquire and manage data from richer sources, model more complex relationships, and provide reference and comparability in broader dimensions. In the final analysis, it is a process of optimizing real time decision-making of management and control activities, and efficient execution of employees being able to assess the state of the processes.

There are three challenges in Industrial Intelligence including "state assessment," "decision-making optimization," and "collaborative implementation." These terms represent the greatest challenges in achieving the above capabilities.

1. State assessment: In order to understand the real-time state of individuals and the environment and how they relate to activities, many of which are not measurable, it is necessary to use modeling to predict the measurable parameters, and more importantly, to accurately evaluate and predict how individuals interact.
2. Decision-making optimization: Decision-making optimization should be based on an accurate and complete understanding of the state, precise analysis, and deduction of the impact of various possible decisions. Full consideration of the

44 3 Definition and Meaning of Industrial AI

trade-offs among multiple objectives must be considered in order to maximize the value of the overall objective.

3. Collaborative implementation: In the process of collaborative implementation, we should consider the hierarchical relationships, timescale, and correlation between decision-making and implementation, and have a pre-determined amount of fault-tolerance.

Knowledge does not exist by itself—it needs to be communicated through a medium. The medium determines the efficiency of knowledge generation, transmission, and application. An example is how, in Japan's manufacturing culture, knowledge is mostly based on using human experience and management systems as the carrier, and is expressed through the training of people and constant improvement of the culture of inheriting knowledge. In German manufacturing culture, knowledge is mostly inherited by focusing on superior production equipment and integrated manufacturing systems, which solidify knowledge into control instructions and logic. With the help of Industrial AI, knowledge can be applied and inherited through data and models, and can help people acquire new knowledge more quickly by constantly finding new relationships and rules from data generated during machine operation.

Future industrial systems will face more uncertainties and changeable environments and systems (environment uncertainty, system uncertainty and task uncertainty). Therefore, smart industrial systems need four basic concepts:

1. Environment-oriented intelligence, which perceives and predicts changes and uncertainties in the environment.
2. State-oriented intelligence, which evaluates and predicts the changes of its own state and the factors influencing risk and performance.
3. Cluster-oriented intelligence, which includes cooperation and collaboration with other individuals and the environment, and helps others learn new knowledge and experience using the activities of individuals.
4. Task-oriented intelligence, which is the transition from "if-then" to "what-if." This requires not only achieving goals, but also includes the knowledge of predicting and managing adverse outcomes.

3.3 GE Predix Successes and Failures

In August 2018, the Wall Street Journal reported that GE was preparing to sell GE Digital's core assets, including Predix, their industrial data platform. As soon as the news came out, there was a view within and outside of the industry that GE's Industrial Internet influence was waning. In fact, the signs of a potential breakup between GE and Predix had begun to appear a year prior. GE was still investing heavily in technology research and development of digital businesses such as Predix, but its revenue had not met expectations. Official data shows that GE Digital had been in a period of decline since 2017, and business value had not achieved the previously expected explosive growth.

3.3 GE Predix Successes and Failures

Of course, the case of GE Predix [2] needs to be looked at dialectically. Its predicament is the result of many considerations, such as enterprise operation, market expansion, and technology deployments—it cannot be generalized. From GE's original plan for Industrial Internet, the direction should have been promising. Key data was obtained from the equipment, and insights were generated through edge analysis. Knowledge decision-making systems were established by using big data analysis and smart algorithms in the cloud platform. Information and decision-making were coordinated among all participants and utilized by the operation and control processes.

This is exactly what an intelligent industrial system should look like, but it is not easy to integrate the personalized business needs of energy, oil drilling, aerospace, and other major industries with a single platform solution. In addition to commercial resistance, a big challenge occurs when technology fails. This is because, although the technical problems encountered in industry are often similar, customer needs are very dependent on personalized usage scenarios. Therefore, in order to truly realize the value of the Industrial Internet, it is more necessary to think realistically and design sturdy systems with the key goal of helping customers to improve their internal core manufacturing capabilities.

On the contrary, there are similar misunderstandings in the development of smart manufacturing in some Chinese enterprises. Promoted by national policy, some enterprises have invested heavily in the "smart" transformation of factories: building Industrial Internet and IoT platforms, adding a large number of sensors to all production lines, creating large databases, and replacing human labor with robots. In the process of this upgrading and transformation, everyone has good intentions, but the actual requirements, needs, and core capacities are often neglected. An analogy would be going to a Golden Corral buffet and only loading up on unhealthy things because they look appealing, while ignoring the healthy and nutritious options.

In my opinion, this is the main problem China faces in intelligent manufacturing. It is also the misunderstanding that the Internet, big data, AI, and other technologies must avoid when being applied to industry. Through the analysis of the present situation and the inherent gaps when dealing with the Industrial Internet demonstrated in the visible-invisible problems quadrant graph in Fig. 3.7, almost all current Industrial Internet platforms focus on using general technology to solve visible problems, and emphasize the tools and technical indicators of the platform, while ignoring the problems themselves.

There are three primary gaps to fill: first, we should be able to solve the visible core pain points in the industrial domain, meaning the problems that enterprises cannot solve. Second, after solving the visible core pain points, we can continue to find the problems and potentials that have yet to be discovered from the historical data of enterprises. Thirdly, by making hidden problems visible and creating new insights, the opportunity space in the six flows of a manufacturing system[1] can be continuously optimized.

---

[1] The "Six Flows" are a concept created by Foxconn founder Terry Gou as a way to conceptualize the manufacturing process. The six flows are the three non-physical streams of information, technology,

Fig. 3.7 Challenges faced by Predix: targeting visible technical problems, but lacking the ability to find problems, solve problems, and gain value in invisible spaces

Another phenomenon that needs to be considered is that, as of 2018, more than 200 Industrial Internet platforms have appeared. Do we really need so many? What are the differences and competitive advantages of these platforms? Technically speaking, we only need one open platform with many domain platforms. A good platform would sell the convenience of solving problems and provide certainty of value, not the selling of tools on the platform. Such a platform should be problem-based, customer-centric, data-supported, tool-based, and value-oriented.

Domain platform should be a problem-focused place, the purpose of the platform being to solve problems. If I am hungry, I can use the Uber Eats app to solve the problem of hunger. I can use Lyft to solve the problem of travel. So, focusing on problem solving, 200 platforms seem like too many. Where is such a platform to solve quality problems? What about platforms to solve energy efficiency and equipment health problems?

GE's Predix's biggest problem was using tools as an end, not as a means to an end. It is true of all kinds of cloud platforms: They promote the machine data to the cloud, but the quality and downtime problems of the machine still exist. If we could put the problem data on the cloud, or embed it within a diagnostic and problem-solving process, then all similar devices could be used as a reference to solve their own problems in a meaningful way.

At this stage, the pain point of the transformation of manufacturing and industrial enterprises towards smart factories is not universal because automation, information technology, IT platforms, big data, and other technologies are not very mature. When

---

and capital, as well as the three physical streams of personnel, material, and processes. Tracking the movement of these six things is a good way to understand what goes into manufacturing.

it is no longer difficult for these technologies to solve visible problems, the most important pain point then will become how to solve and avoid them. The invisible problems that will be encountered include deterioration of equipment performance, loss of machine accuracy, wear and tear of parts, and waste of resources. Visible problems are often caused by accumulations of these invisible factors. Like an iceberg, the visible problem is only a small, but visible part of the entire picture. The function of industrial big data and AI is to obtain deep insight into invisible problems, and create worry-free manufacturing by predicting the invisible characteristics of a production system.

Following this idea, in Fig. 3.8 we can further divide the opportunity space of Industrial Intelligence transformation into four parts.

The first opportunity space comes from solving the visible problems in a production system. There are still issues that many manufacturers need to better acknowledge and address in this space, such as long product delivery cycles, excessive environmental pollution, and waste of resources. In order to solve problems in this opportunity space, we need continuous Lean improvement of the production system and continuous improvement of system standardization.

The second space is avoiding visible problems, which can be accomplished by mining data to build new knowledge to address the causes of deeper problems and improve the original production system and products. For example, a factory can analyze the causes of equipment failures throughout the year using Six Sigma practices, find the most frequent failures and most influential factors, and then establish the corresponding prediction model to intervene in advance.

The third space is concerned with building dominance over invisible problems which should begin by using innovative methods and technologies to find solutions

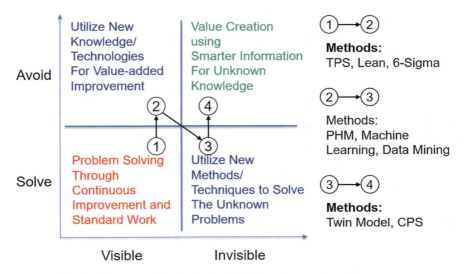

Fig. 3.8 Opportunity space and path of industrial information transformation

and by enhancing the theoretical limits of modern manufacturing systems. More in-depth data, as well as relevance and causality mining, are needed. Through the establishment of these relationships, invisible processes can be quantified. For example, the relationship between process parameters and quality is to achieve predictive quality management through comprehensive modeling and analysis of process parameters, operating environment, and equipment status.

The fourth space is finding and satisfying the invisible value gap and avoiding the influence of invisible factors. This opportunity space requires the use of big data analysis to expand the scale of relationship building to create higher dimensional knowledge, through which closed-loop integration of the manufacturing industry chain and optimization of the upstream design of the system to avoid invisible problems can be achieved. For example, we can now assess the health degradation of wind turbine through Industrial Intelligence technology. In the future, we will be able to associate the knowledge generated in this process with the structural design and control logic of the fan and design the control logic of the fan so that it can operate independently and reliably. In health mode, the optimal control parameters can be used to achieve optimal performance. Industrial Intelligence has already reached this point.

At this stage, the vast majority of manufacturing enterprises are concerned about improving the first and second spaces. However, Industrial Intelligence technology should be applied to solve invisible problems in the third and fourth spaces. Only then can we achieve breakthroughs in production efficiency and the product competitiveness of manufacturing systems, and finally achieve the smart transformation of industrial systems.

3.4 Technical Elements of Industrial AI: Data, Analytics, Platform, Operations, and Human-Machine Technologies

Currently, small and medium-sized enterprises are still in the initial stage of Industrial AI, and their structure, methods, and challenges must be clearly defined as within the framework of industrial implementation. Fig. 3.9 shows the system framework of the Industrial AI Center (IAI Center, https://www.iaicenter.com) which applies Systematic, Speedy, and Sustainable AI to business. This system covers the basic elements of Industrial AI and provides guidelines for better understanding and implementation of the ecosystem.

An Industrial AI System defines the strategic requirements, challenges, technologies, and methods that need to be considered in AI systems applied to industry. This ecosystem divides according to the fields of industrial manufacturing, such as embedded AI equipment, flexible manufacturing systems, worry-free logistics, and predictive energy systems. It also proposes common requirements in these fields

3.4 Technical Elements of Industrial AI: Data, Analytics …

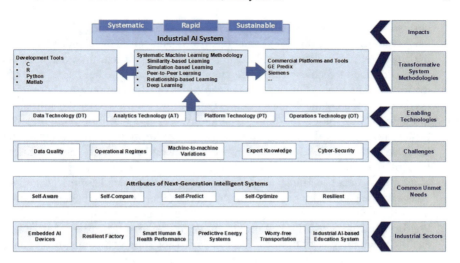

Fig. 3.9 Framework and technical elements of an industrial AI system [3]

such as self-awareness, self-comparison, self-prediction, self-optimization, and self-resilience. Among them, self- resilience refers to whether the system can handle equipment failure or errors, and quickly restore its normal working state. If the redundancy and flexibility of the system are predefined, the system can quickly resume normal operations by rescheduling and deploying tasks when serious disturbances occur.

Industrial AI frameworks use Data Technology, Analytic Technology, Platform Technology, Operations Technology, and Human-Machine Technology to solve the above challenges. These five technical elements are briefly described below:

1. Data Technology (DT): In the manufacturing domain, each step generates unique data every time it is run. Cumulatively, these unique steps in individual processes can generate gigabytes of valuable data. Data can be generated at either the component level, the machine level or the shop-floor level, and can be broadly divided as structured and unstructured data. Solving the 3B problems in industrial data, broken data, bad data, and the background of data, as applied to the CPS is very important. In addition to achieving the unified collection of heterogeneous data, we also need to achieve the automatic extraction of effective data and improve the standardization of data acquisition modelling. Also, when managing data, we should pay attention to data synchronization. For example, in a manufacturing production line, equipment parameter data needs to be associated with a product throughout the entire process.
2. Analytic Technology (AT): Sensors and data acquisition systems in the data technology implementation phase would provide abundant data that needs to be analyzed to extract meaningful information from it. Analytics plays the role of the interface between the physical world and the cyber world. Analytic technologies

correspond to the conversion layer in CPS architecture, including pipeline analysis at the edge (fog computing) and higher-level analysis in the cloud. Although the edge end is closer to the industrial site and has good real-time data analysis, it is difficult to achieve centralized model training and prediction because of the limited data processing and storage capacities of edge computers. Therefore, according to the characteristics of edge computing, we need to push for breakthroughs in flow-based reasoning technology, and use real-time data streams to iteratively refine the relevance of data to achieve model self-optimization. Because of this small sample data, new modeling and analysis methods need to be explored, such as semi-supervised learning, time-machine based state modeling, peer-to-peer learning, adaptive learning, and transfer learning. Another area with the potential for breakthrough thinking, there is also a tendency to use high-speed computers at the edge, which will be introduced in later chapters.

3. Platform Technology (PT): Platform technologies serve as the carrier of all other enabling technologies of Industrial AI systems. They support the function of connection layer, conversion layer, and cyber layer of a CPS architecture. Similarly, for edge computing platforms, we need to improve the ability of signal acquisition and computing through breakthroughs in hardware technology as well as expanding the collaborative capability of edge platforms and supporting self-organization and self-configuration of equipment in the production line. With this interconnection comes a greater need for network security. Hardware architecture and software mechanisms need to ensure the security of system access, especially after realizeing the feeback control of edge devices. For cloud platforms, we should focus on building components of life cycle management model, assisting uncertainty management model, and continuous self-learning model of Industrial Intelligence. Overall, platform technologies can be categorized into three level of services: Infrastructure as a Service, Cloud as a Service, and Solutions as a Service.

4. Operations Technology (OT): This technology is realized through the cognition layer in the CPS architecture. Operations technology, in conjunction with the above three technologies, aims to optimize production throughout manufacturing operations. It depends on the upgrading of operational management methods. This includes how to effectively transform the knowledge gained from the prediction model into operational maintenance and management decision-making and realization of the transformation from experience-driven production to data-driven production.

5. Human-Machine Technology (HT): HT corresponds to the configuration layer of CPS architecture. Industrial Intelligence will greatly affect how manufacturing systems interact with people. It can be expected that, in the future, the amount of information processed at all times in factories will increase dramatically. How to help producers obtain the most effective and relevant information in the most intuitive and seamless way will become a major challenge. Therefore, it is necessary to explore the human-machine interaction within the industrial domain. For example, intelligent assistants would provide smart reminders to station personnel, or virtual reality (VR) and augmented reality (AR) could assist in the

3.4 Technical Elements of Industrial AI: Data, Analytics … 51

remote diagnosis of technology. Human-machine technology will help Industrial Intelligence technology better integrate into production and empower producers.

These technological elements will not exist in isolation, but need to be integrated into a system in order to play a real role. The 5C architecture of CPS is the core functional framework that integrates the above technologies. This also means that future development of Industrial Intelligence technology will still be an interdisciplinary process which requires the combination of multidisciplinary domain knowledge. Among the five core technologies of Industrial AI, the analytic technology that can deeply integrate with business value is the soul of Industrial Intelligence. Data technology and platform technology are the necessary conditions for industrial intelligent landing and the carriers of intelligence. This is because efficient data connection and mature platform technologies are the prerequisite for smart systems and are necessary for Industrial AI systems.

Finally, operational technology is the key to value creation. Based on the results of intelligent analytic technology, operational technology provides decision-making suggestions and behavioral recommendations for users through optimization, which is also an important part of an AI system.

According to the above five technologies, each company will have different characteristics and development paths according to their domain. Foxconn, for example, is the largest manufacturer of electronic products in the world, with more than 80,000 industrial robots, 1800 surface mount technology (SMT) production lines, 175,000 CNC and die processing machines, and more than 5,000 kinds of testing equipment. Therefore, within the Industrial AI field, the superior competitive advantages of Foxconn include experience and expertise in factory environments, within the manufacturing field, with equipment maintenance, computers, facilities, and huge amounts of industrial data, and in fields related to other industrial mechanisms. When all of the equipment in the group is connected to the network, it can achieve smart perception, smart analysis, and smart networks, and fulfill the goals of improving quality, increasing efficiency, reducing costs, and reducing inventory.

Foxconn implements similarity-based learning, simulation-based learning, peer-to-peer based learning, relation-based learning, and deep learning by using tools on the Foxconn Industrial Internet Cloud platform. Collected data is trained through systematic machine learning to form an AI model. Then, the model is deployed to a cloud platform consisting of Foxconn Industrial Internet Cloud, CorePro, Nadder, and other Industrial Internet products developed by Foxconn.

Currently, the Foxconn cloud platform has been widely used internally. CorePro supports the collection, parsing, and storage of data from serial ports, documents, software, or anything communicating using standard protocols. Micro Cloud platform is used to build the application system across the edge, core, infrastructure, platform, and software layer, connecting the equipment, software, workshops, enterprises, managers, and decision-makers.

However, in order to facilitate the continuous flow of data through classification, analysis, and cognitive improvement that achieves seamless integration between domains and their data, Foxconn developed a new concept: Fog AI. Looking at

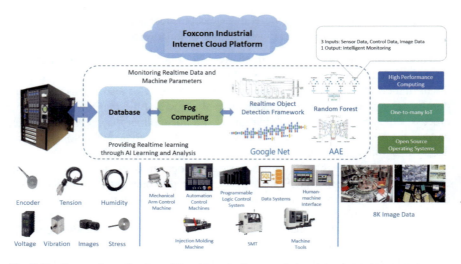

Fig. 3.10 Foxconn's application of Fog AI on the Foxconn Industrial Internet Cloud platform

Fig. 3.10, Foxconn's Fog AI technology utilizes the integration of high-performance computers with the server-level to collect data. After establishing the data model, Fog AI can truly achieve real-time prediction and monitoring, accurately control the production process, and allow for more stable, accurate, and faster responses from equipment. The following chapters will also introduce how Foxconn uses Industrial Intelligence for nozzle suction and tool lifetime prediction, all with the goal of improving plant efficiency and reducing inventory costs.

3.5 CPS: An Architecture for Integrating the 5 Technological Elements of Industrial Intelligence

In the previous section, we introduced the elements that support an Industrial Intelligence system. We need to be able to integrate these technologies into a system architecture that can bring out latent potentialities. The 5C architecture of a CPS is the core framework to integrate the above five technological elements. The German Industry 4.0 strategy and the American CPS plan each regard CPS as the core technology of implementation and use it in their strategic goals. So, what is CPS?

CPS is a smart industrial technology system to address the whole process of data collection, aggregation, analysis, sorting, prediction, decision-making, and distribution, and can analyze industrial data in a streamlined and real-time way. During analysis, the characteristics and requirements of logic, relationships between processes, and business activities are each fully considered. Therefore, CPS is the core of building an intelligent system through the analysis of large industrial data.

3.5 CPS: An Architecture for Integrating the 5 Technological …

In the process of production, lots of data is generated. The essence of Industrial Intelligence is to extract logical relations and knowledge from production data by using big data and AI, and to formulate decisions and appropriate actions to avoid invisible problems and transmit manufacturing knowledge.

For example, for a machine tool that processes molds, the worker is controlling the machine, and the cutter is cutting. The equipment is generating a lot of vibration signals. In the cutting process, the tool can be stabilized at the optimal speed and angle without resonance and flutter using intelligence, and it can consider the processing characteristics of the material to ensure the final product quality. If the tool is worn out after cutting for a period of time, the machine should be able to find out by itself and stop automatically before breakage occurs. Nearby workers should be notified that they need to replace the tool, but newcomers to the factory might not know how to replace the tool. In this case, the machine should be able to demonstrate this process to prevent unnecessary errors. At this point, the industrial domain is practically intelligent.

To build a smart factory, it necessitates a large amount of data and building a closed-loop system to transform data into knowledge and then into execution. In past work, I have put forward the 5C architecture of CPS as the functional framework to realize this closed-loop process. 5C represents connection, conversion, cyber, cognition and configuration. Their more specific meanings is below (Fig. 3.11).

Connection: The connection layer is about smart sensing. The quality and comprehensiveness of data are guaranteed from information sources, collection methods, and management methods. The data environment foundation supporting CPS is established. In addition to the establishment of interconnected environment and data acquisition channels, another core of intelligent sensing is to independently select and focus on data acquisition according to the objectives of activities and needs of information analysis.

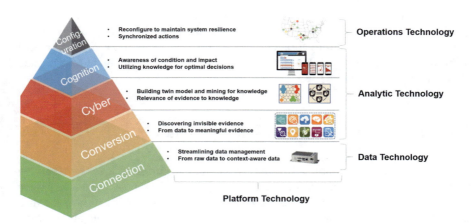

Fig. 3.11 The '5C' architecture for cyber-physical system (CPS) [4]

Conversion: The conversion layer transforms low-value-density data into high-value-density information which can extract features, filter, classify, and prioritize the data, ensuring interpretability and allowing for data segmentation, decomposition, classification, and analysis.

Cyber: They cyber layer focuses on the cyberspace modeling in the network environment. A modeling and analysis environment can guide the entity space, including accurate synchronization, association modeling, records of change, analysis, and prediction.

Cognition: In complex environments and conditions with multi-dimensional data, evaluation and prediction of multi-source data is carried out according to different evaluation needs in order to recognize the operational patterns of systems and the relationships between objects, environments, and activities. Data visualization and decision optimization tools provide users with the support to make decisions.

Configuration: According to the objective of an activity and the analysis parameters in the cognitive layer, the operational decision is optimized and the optimization results are synchronized with the execution of the system to ensure the timeliness of information utilization and system operations synergy.

The reason why CPS can improve the core competencies of manufacturing is because it integrates the ever-expanding growth of computing power within the information world and the manufacturing capacity of the physical world into an integrated system. It breaks through the limitations of the traditional production system and greatly develops it. Therefore, if we rethink the technology of the Industrial Internet and AI from the perspective of CPS 5C architecture, we will find that smart manufacturing is not simply equivalent to production process monitoring or the application of deep learning. At most, these are components of CPS. Fig. 3.12 illustrates how four enabling technologies of Industrial AI (DT, AT, PT and OT) act as enablers for achieving success in the Connection, Conversion, Cyber, Cognition and Configuration steps of 5C architecture.

In order to fully integrate these smart technologies in the industrial domain to solve industrial problems, it is necessary to systematically and structurally establish the link between the information and physical worlds, to find the most important impact parameters for problem solving rather than blindly accumulating data, and finally to form closed-looped systems.

3.6 Industrial AI: Categories of Algorithms

Functionally, AI algorithms can be divided into the categories of supervised learning and unsupervised learning according to whether or not the labels of data are known during training. As shown in Fig. 3.13, supervised learning refers to how the data input during model training includes input objects (usually a set of feature vectors) and their expected values (discrete or continuous values). According to the training, supervised learning aims at generating the relationship between input feature vectors and expected values, generating a mathematical model with an inferential prediction

3.6 Industrial AI: Categories of Algorithms

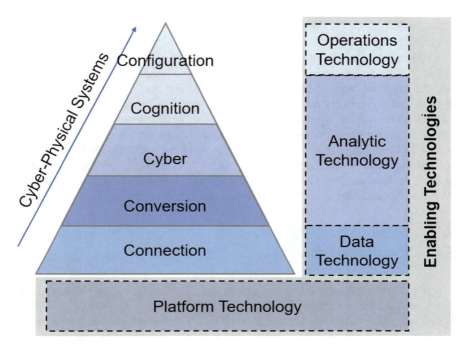

Fig. 3.12 Enabling technologies for realization of CPS in manufacturing

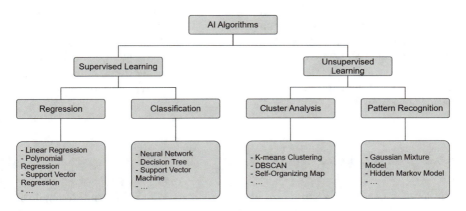

Fig. 3.13 Categories of Industrial AI algorithms

function for mapping new input objects. During training, supervised learning can be classified into regression or classification, according to the continuity of expected values.

In regression algorithms, the training labels are continuous, and the trained model can infer the corresponding expected value according to the new input object. In the classification algorithm, the training labels are discrete values, and the trained model can classify the unlabeled new input objects into corresponding categories.

Unlike supervised learning, training data in unsupervised learning does not contain labels. The process of learning is not to find the relationship between the input object and the expected labels, but to recognize the patterns amongst the input objects. Typical unsupervised learning processes include clustering algorithms and statistical estimation. Clustering algorithms aim at grouping similar input objects together to form distinct categories, while statistical estimation algorithms use the principle of probability and statistics to describe the input object as a distribution function with specific parameters or to estimate the correlation of a time series between input data. In the following section, we will introduce several typical learning models for each AI algorithm and their applications in industrial data analysis.

3.6.1 Regression Algorithms

In regression algorithms, the input feature vector $\vec{x}$ and the label value y in the training dataset are used to establish an estimation function so that $y = f(x)$ or $y \approx f(x)$. When such an estimation function or model is established, we can estimate the expected value y of the new input feature vector x. In industrial scenarios, such algorithms are often used for virtual metrology or system health assessment. Virtual metrology means that the quality of the product or stability of the production process on the line can be estimated by the data collected in the production process, without additional measurement or detection processes. The health assessment of a system usually uses the input and output of the system to establish a model. The system output obtained by the comparative measurement and the system output estimated by the model are used to achieve the monitoring. The regression algorithms commonly used in the industrial domain are linear regression, polynomial regression, and support vector regression.

3.6.2 Classification Algorithms

Similar to regression algorithms, the classification algorithm also wants to obtain an estimation function $f(\cdot)$, but the result of the estimation is not a continuous value; it is a discrete value. Possible discrete values form a set in which each element is a possible category. The task of the classification algorithm is to construct such a classifier, $f(\cdot)$, so that there is only one class corresponding to the input feature vector.

In industrial data analysis, classification algorithms are often used to diagnose faults or trace their cause. Classification models are established by using historical data and the corresponding labels. In the process of analyzing the newly collected data, the classification model can estimate the possible failure modes or the important factors that cause certain types of faults. The classification algorithms commonly used in the industrial domain mainly include support vector machines, neural networks, and decision trees.

3.6.3 Clustering Algorithms

Clustering algorithms cluster the input feature vectors unsupervised on the basis of similarity, which makes the similarity within clusters larger than that between clusters. Because input data has no label information in the clustering process, the clustering only estimates the dataset and recognizes the pattern. Clustering results often need expert experience to explain the characteristics of each cluster. In industrial scenarios, clustering algorithms are mainly used to identify different working conditions and to evaluate the health of the system.

When the working conditions of the system are complex, it is necessary to identify these different conditions before analyzing the data in order to establish corresponding analysis methods for different working conditions. In the absence of labeled historical data, we can first cluster the data with its health status to identify patterns. The health status of the system can be obtained by comparing the newly collected data with the identified health patterns. Common clustering algorithms include K-means cluster, DBSCAN, and self-organizing maps.

3.6.4 Statistical Estimation Algorithms

Statistical estimation is another standard unsupervised pattern recognition algorithm. It uses the principles of statistics and probability theory to identify the potential statistical distribution form of the input dataset for estimation. Datasets can be represented as a combination of one or more distribution functions—the relationship of a time series between each input feature vector of data can also be represented by probability and state transition functions.

Statistical estimation can be used to represent the current state of the system, or the entire decay process of the system, and can characterize the distribution of data under different fault modes and degrees. Based on estimated data distribution patterns, users can deepen their understanding and quantify the risks of equipment operation. Common statistical estimation algorithms are the hidden Markov model and the Gaussian mixed model.

3.7 Industrial AI Algorithms: Selection and Application

Although there are no fundamental differences between industrial and non-industrial AI algorithms, we must understand what makes an algorithm suitable for application in the industrial domain. There is no general algorithm for industrial data. Based on the introductions of algorithms in the previous section, this section will give guidance to selecting algorithms according to specific industrial applications.

In the process of selecting algorithms, we first consider the following factors: the sources of data, working conditions, clustering characteristics, fault modes, and expertise. As shown in Table 3.1, we suggest the corresponding algorithm selection strategies according to the above factors. When the data source is insufficient—for example if there is only a controller signal with low sampling frequency—augmented learning can be used. This is when one uses third-party data acquisition and analysis equipment to obtain more signals and carry out targeted analysis. When there are enough data sources, deep learning can be considered to fully mine the implicit information in the data.

Looking at working conditions when the operation conditions of the system are complex, such as the various processing procedures of CNC machine tools or continuous alteration of heating and refrigeration in HVAC (heating, ventilation and air conditioning) systems, we give priority to clustering algorithms in the selection of working conditions. We can also do Fixed Cycle Feature Tests (FCFT) where the system is run repeatedly under fixed conditions during a fixed period. Once the corresponding data is collected, then the performance and decline of the system can be estimated according to the data.

The third consideration is clustering characteristics. If the object of analysis is clustered and there are many identical or similar devices in the monitoring state, then we can consider applying similarity learning or broad learning to analyze the relationship between the device and its adjacent devices, or use the information of these adjacent devices to estimate the status of the device. In addition, when the historical data of the system contains a large amount of fault mode information and

Table 3.1 Appropriate algorithms for industrial applications

	Usability	
	Low	High
Data sources	Augmented learning	Deep learning
Working conditions	General AI algorithms	Clustering algorithm, fixed working conditions
Clustering characteristics	General AI algorithms	Similarity learning, board learning
Fault mode history	Unsupervised/semi-supervised learning	Supervised learning
Expert experience	Deep learning, ensemble learning	Fuzzy logic

3.7 Industrial AI Algorithms: Selection and Application

corresponding historical data, we can use supervised learning to model the health and fault data of the historical data and analyze the collected signals. However, when the historical data of the system contains less fault mode information, we usually use unsupervised or semi-supervised learning methods only for health data modeling, and enrich the model according to needs in the monitoring process.

Expert experience is another factor to consider. Our knowledge of a system often helps us better understand that data and develop targeted data acquisition strategies. When we have a good understanding of the system's operation status, common failure modes, and mechanisms of occurrence, we often use AI algorithms such as fuzzy logic to introduce expert experience into the analysis model. When the expert experience of the monitoring object is scarce, we need to consider using deep learning algorithms to analyze the original data and understand the data patterns with the help of the algorithm. On the other hand, we often use ensemble learning to train several AI models at the same time, and give a comprehensive output based on the results of each model.

After determining the categories of algorithms, we must consider which specific algorithms we want to use in this category. As shown in Table 3.1, we consider the complexity and uncertainty of the system in two axes, and divide the object and data into four quadrants according to these two axes, labeled A, B, C, and D.

In area A, the complexity and uncertainty are low. For example, a single CNC machine tool uses the same spindle speed to produce a single product. In this case, we only need to consider general AI algorithms to use during data analysis to obtain good results.

In area B, the complexity is low while the uncertainty is high. The data in the monitoring process is vulnerable to noise or environmental factors, and the algorithm we choose needs to take this into consideration. If necessary, we can estimate the uncertainty so that the analysis model can adapt to it.

In area C, the uncertainty is low, but the complexity is high. This often happens due to complex working conditions, such as when cutting various materials, the different shapes of tools, and different branding needed on items. Hence, the selected algorithm must have high robustness and self-learning so it can adapt to different working conditions.

In area D, both complexity and uncertainty are high. In this case it is difficult to model the system data. We can use data to describe the state and behavior of the system by estimating its probability, such as by using the naïve Bayesian algorithm or by introducing expert experience to build a relational model.

As we consider the complexity and uncertainty of a system, we also need to consider the computational complexity of the algorithm and accuracy of the final results when selecting an algorithm. These two dimensions are also important in practical engineering applications. After defining the scope of several candidate algorithms, we can narrow down the scope through visualizations like Fig. 3.14. In Fig. 3.15, we also divide the space into four quadrants according to the design requirements of result accuracy and computational speed.

In area A, the accuracy of the results and the rate of computational speed are not very high, and the algorithm has no further requirements.

Fig. 3.14 Four-quadrant diagram of system complexity and uncertainty

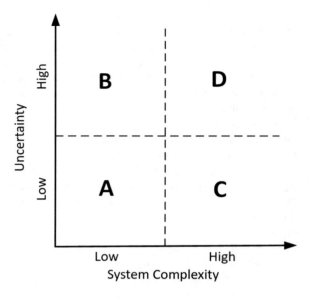

Fig. 3.15 Four-quadrant diagram of accuracy and computational speed

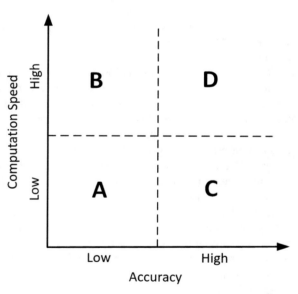

In area B, the accuracy of the results is low, but the computational speed is high. This area usually occurs in scenarios where the analysis results are displayed in real-time. For example, in semiconductor fabrication, the processing results of each wafer need to be evaluated in order to adjust parameters in real-time. Some fast-convergent algorithms which are insensitive to the dimensionality of the data can be used including linear regression, logical regression, and so on.

In area C, the accuracy requirements are higher but the computational speed requirements are lower. Most application scenarios can be classified in this area. We hope to know the operation and health information of the system in the form of reports, but we need high result accuracy. Thus, we can choose algorithms like deep learning or ensemble learning to make the best usage of each model and utilize the advantages of each model in the data at the expense of computational speed in order to obtain maximum performance.

Region D indicates that there are high requirements for computational speed and accuracy. This is generally the case for situations where real-time control of a system is needed, and data analysis is quickly utilized. An example is routing multiple ships in real-time according to weather and ocean currents to avoid collisions. In this region, some rule-based algorithms based on expert experience often perform well such as the combination of fuzzy logic and genetic algorithms.

References

1. FANUC (2019). *FANUC FIELD system ZDT-Zero Down Time*. Available from: https://www.fanucamerica.com/products/robots/zdt-zero-down-time
2. Peter C. Evans and Marco Annunziata (2016). *Industrial Internet: Pushing the Boundaries of Minds and Machines*. Available from: http://files.gereports.com/wp-content/uploads/2012/11/ge-industrial-internet-vision-paper.pdf
3. Lee, J., Davari, H., Singh, J., & Pandhare, V. (2018). Industrial Artificial Intelligence for industry 4.0-based manufacturing systems. *Manufacturing letters, 18*, 20–23
4. Lee, J., Bagheri, B., & Kao, H. A. (2015). A cyber-physical systems architecture for industry 4.0-based manufacturing systems. *Manufacturing letters, 3*, 18–23

Chapter 4
Killer Applications of Industrial AI

4.1 Application Scenario Types for Industrial AI

According to Foxconn Industrial Internet (Fii), the application scenarios of Industrial Internet can be divided into four types: Factory, Field, Fleet and Facility (4F). These four application scenarios also have different tendencies towards value. The factory scenario focuses on the product and the production process and pays more attention to improving quality, increasing efficiency, reducing costs, and maximizing savings. The field scenario pays more attention to equipment status maintenance and personnel-task matching optimization. Fleet operation pays more attention to dynamic operations and tasks. Collaborative optimization is carried out in dynamic environments and values cooperative optimization, continuous optimization of the turnover rate of use, improvement in the performance of tasks, and reduction in the cost of integrated operation and maintenance. Facility is more focused on continuous and stable operation, automation, and energy efficiency optimization.

These application scenarios can also overlap. For example, in the application of wind power generation, there are two attributes: field and fleet, which also have the application requirements of these two scenarios. In the factory management system, the intelligent FMCS (Facility Monitoring and Control System) should consider both the application requirements of equipment and facilities and the requirements of the factory production scenario (constant temperature, constant pressure, cooling, compressed air, etc.). Therefore, the main starting point of dividing the 4F's are the challenges, requirements, and commonalities in the application scenario, which can be used for reference in different objects and industrial types.

From the perspective of algorithm and analytic technology, Industrial AI algorithms can be divided into the following application functions:

Classification classifies new input data according to a set of training data. Its main task is to identify the label information of testing inputs, such as trucks, cars, products subject to quality inspection on production lines, etc.

Continuous estimation is the sequential valuing of new input data according to the training data. It is common in predictive tasks, such as predicting spare parts

© Shanghai Jiao Tong University Press 2020
J. Lee, *Industrial AI*,
https://doi.org/10.1007/978-981-15-2144-7_4

63

demand based on various dimensions of data, predicting product quality (virtual metrology) based on process parameters, etc.

Clustering is the creation of a single cluster in a system based on task data. The cases are consumer preferences based on personal data.

Operational optimization is when the system produces a set of output function optimization results as a specific goal according to the task. The cases include scheduling optimization, maintenance scheduling optimization, location optimization, unmanned vehicle scheduling optimization, etc.

Anomaly detection, which judges whether the input data is abnormal according to the training data/historical correlation, can be considered as a sub-category of the classification function; application cases include multivariate process anomaly detection, equipment health warning, network intrusion detection, etc.

Diagnostics is a common problem in information retrieval and fault diagnosis, that is, presenting results according to certain ranking criteria based on retrieval needs. Cases include providing product purchase recommendations, anomaly detection recommendations when defective products occur, etc.

Recommendations, i.e. recommendations for a specific activity goal based on training data, such as maintenance plan recommendations.

Prognostics is the prediction of possible anomalies in the future by continuously evaluating device parameters, including the time of occurrence, failure mode, and impact.

Parameter optimization, through the establishment of a model between multiple control parameters and the impact equation on the optimization objectives, combined with optimization algorithm, is the dynamic optimization of the combination of multiple control parameters, including boiler combustion optimization, heat treatment process parameters optimization, and so on.

The above are just different analytic functions. They do not have clear scenario attributes. In other words, the requirements in various scenarios can be abstracted into the above categories of problems and objectives, and the most appropriate solution can be chosen according to the individualized objectives and constraints in different scenarios. Problem-solving tools can meet the needs of most scenarios.

Another categorization method is based on the application tasks. In the series of research reports on data-driven transformation and Industrial Internet (Fig. 4.1) published by a Chinese research center, the business functions for Industrial Internet can be divided into three categories: (1) equipment and product management, (2) business and operation optimization, and (3) service and business model innovation. We believe that this categorization is also applicable to the business function division for Industrial AI. From these business function scenarios, we can see that the value of Industrial AI for enterprises mainly includes cost savings, increased efficiency, increased product and service value, and exploring innovative business models. Around these four core business values, Industrial AI will produce "killer applications" in many business areas, and gradually produce composite scenarios, application collaboration, and cross-domain Industrial Intelligence application platforms.

4.2 What Will Become the "Killer Applications" of Industrial AI?

4.2.1 Predictive Maintenance of Equipment

In the past, due to the lack of accurate methods for determining when an equipment failure would occur, equipment maintenance operators faced a dilemma: whether to maximize their use time at the risk of failure or to replace normal components in advance to ensure the maximum reliability of equipment operation. According to Deloitte's report, "Predictive Maintenance and the Smart Factory" published in 2018 [1], unreasonable maintenance strategies lead to between a 5% and 20% reduction in factory capacity, and the losses caused by an industrial enterprises' unexpected shutdown can be as high as $50 billion per year.

Traditional systems have been able to analyze sensor time series data, including temperature, vibration, etc., to detect faults and predict maintenance (to predict the remaining useful life of components). But deep learning takes this function to a new level: data can be stratified to analyze massive, high-dimensional sensor data, including images, audio, and other forms of sensor data. Some previously unusable, low-quality data (from cheap microphones and cameras) can also be used by deep learning. In the case studies, this maintenance prediction (remote airborne diagnostic technology) based on AI technology can help enterprises reduce downtime, formulate planned interventions, increase production, and reduce operating costs.

In short, predictive maintenance can quickly collect data from key equipment sensors, the enterprise resource planning system (ERP), the computer maintenance management system (CMMS), and other systems. Intelligent factory management systems combine data with advanced prediction module and analysis tools to predict equipment failure and deal with it. This can help maintenance personnel find the root cause of the problem.

The value of predictive maintenance is based on the prediction of remaining useful life. In the maintenance opportunity window, the maintenance strategy and scheduling plan with the optimized cost are selected. At the same time, the maintenance requirements for all equipment are considered comprehensively, and the global optimal maintenance plan is formulated. In short, it is used to determine uncertain information and save costs while improving efficiency for customers.

According to a survey of 153 mechanical engineering operators conducted by Roland Berger in collaboration with Hannover Messe, 81% of the companies surveyed had planned predictive maintenance [2]. The main technical challenge for predictive maintenance is a combination of trying to bridge different disciplines and use of multiple technologies. Furthermore, the different industries in which this will be used each have their own environmental and objective particularities. In addition to some common problems (e.g. fault prediction of rotating machinery), a practical challenge that is hard to avoid is that it is difficult to accumulate knowledge in a short time for a sub-area, let alone to complete the whole delivery of the project in a few months.

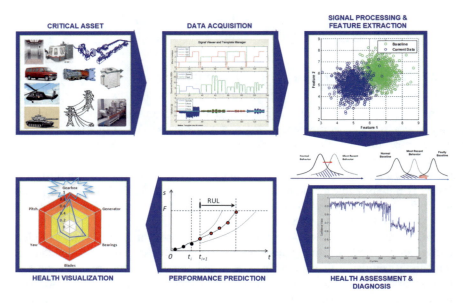

Fig. 4.1 Engineering data analysis process with the goal of achieving predictive maintenance

In addition to the above-mentioned investigation and study of relevant cases, a systematic method for problem abstraction is also needed. In the book "Industrial Big Data: The Revolutionary Transformation and Value Creation in INDUSTRY 4.0 Era" [3], we introduced a systematic methodology for engineering data processing with prognostics and health management (PHM) as the core technology. This methodology includes six main steps: data acquisition, signal processing, feature extraction, health assessment, health prediction, and visualization (see Fig. 4.1). Here, we introduce a method to abstract business objectives and to design modeling paths in the early stage of research and illustrate them with two application cases in Foxconn's unmanned factory.

The visible versus invisible quadrant graph (Fig. 4.2) is used to express the path for the predictive maintenance of equipment. On the basis of real-time monitoring, precision management and fault diagnosis, the quantitative evaluation of the fault process and performance degradation is further realized allowing for early intervention and avoidance of system failures. Faults are visible problems, so the causes of equipment faults, as well as the performance differences between equipment and other factors are invisible problems. So, we need to understand and solve these invisible problems through the analysis and mining of large data and peer-to-peer analysis. When these two goals are achieved, worry-free factories with zero downtime can be realized by integrating information with humans and systems.

4.2 What Will Become the "Killer Applications" ... 67

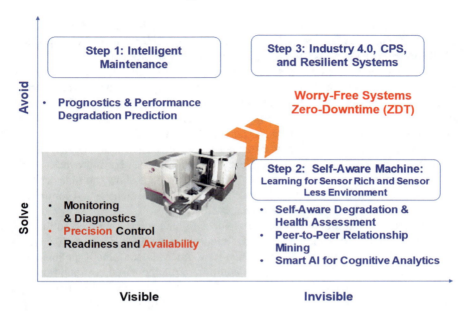

Fig. 4.2 A worry-free factory path for zero downtime

4.2.1.1 Case Study 1: PHM for the Suction Nozzle in Surface Mount Technology Production Line

The first case study is the prognostics and health management (PHM) for the printer suction nozzle in the surface-mount technology (SMT) production line. SMT is a very mature and highly automated process in the factory in which electronic components are soldered onto the surface of printed circuit boards (PCBs). As shown in Fig. 4.3, the process includes printing solder paste through a printer onto a PCB, then to use an integrated circuit (IC) mounter to make pieces (such as resistors, capacitors, diodes, transistors, integrated circuits, etc.) and then use hot air from a reflow oven plant to

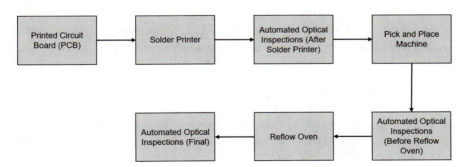

Fig. 4.3 Surface-mount technology process

make tin paste melt, which will combine electronic parts and the PCB, completing the assembly and welding technology, and then through automatic optical inspection (AOI) optical detection instrument, control the quality of production.

The IC mounter in the SMT process uses vacuum suction to move electronic parts. It transmits suction through filters to pick up electronic parts, such as ball grid array (BGA) IC, connector etc. Then it places these electronic parts in the correct position of the circuit board. However, the remaining useful life and usage degree of the filter element of a suction nozzle will affect the production yield, suction maintenance, replacement time, and throwing cost. If the health status of a suction nozzle can be predicted in advance, the process efficiency and stability can be improved as well as the competitiveness of the company. This case introduces how the Industrial AI framework is applied to the mounting machine to effectively manage and control the life and health cycle of the suction nozzle.

Generally, the traditional suction nozzle doesn't have a label number, so there is no way to collect and track the use of the suction nozzle filter element effectively. It is necessary to inspect dirty filters manually in order to determine if maintenance or replacement should be carried out. In addition to the increased cost and time for manual inspections, the judgment criteria for inspectors is different, which results in a missed detection rate of up to 10%. Therefore, in the early stages of the project, the team tried to find out the key factors affecting the life of the suction nozzle through data acquisition at the equipment production end. By collecting the data from 57 h of production, it was inferred that the suction times for the suction nozzle are strongly correlated to the vacuum value. The vacuum value can be regarded as a key reason for failure. And the biggest factor affecting the vacuum value of the suction nozzle is the cleanliness of the filter element of the suction nozzle. With the data as evidence, the team further proposed a framework to establish a suction nozzle health model.

If we can collect real-time data (suction times, throwing rate, yield of fittings) and establish a model to evaluate the dirty area of the filter core and monitor the health cycle for the suction nozzle, and clean or replace filter cores automatically before the aging degree of the suction nozzle reaches a critical level, then we can achieve the goal of both reducing the failure and missed detection rates, shortening the manual operation time for workers as the first step.

In DT data acquisition, the suction nozzle is provided with a separate ID using a radium engraved barcode. Then by linking IoT technology with the machine, data for the suction nozzle continuing to be used to the vacuum door threshold or suction nozzle failure of the equipment are collected, and the correlation of the failure zone is obtained by using the distribution profile.

After the key effective data acquisition, intelligent models are trained to automatically identify the dirty area of the filter element, and then to prolong the life of the suction nozzle. Figure 4.4 shows the transition for a filter in the initial clean state to a dirty filter in the final state. At present, this model has been continuously optimized and improved by Foxconn engineers, and has effectively prolonged the life of the suction nozzles used in their production processes, to the point that each suction nozzle has its own independent health model. As shown in Fig. 4.5 the developed App can promptly respond to the sudden abnormalities of the blocked suction nozzle

4.2 What Will Become the "Killer Applications" ... 69

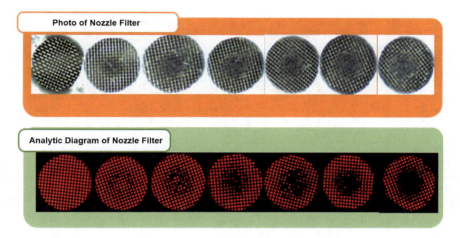

Fig. 4.4 Identifying the degree of dirtiness of the nozzle filter

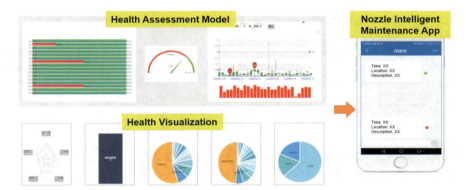

Fig. 4.5 Nozzle predictive maintenance App

and inform the relevant personnel, as well as predict the remaining useful life of the suction nozzle.

By predicting the life cycle of the suction nozzle and replacing it automatically, the maintenance time of the suction nozzle can be effectively shortened, the cost of wasted material can be reduced by up to 66%, and the inventory of the suction nozzle can be reduced by 64%.

The technical framework of the suction nozzle health cycle prediction includes data technology (DT), analytic technology (AT), platform technology (PT), operations technology (OT), and other fields, as shown in Table 4.1.

Table 4.1 Technical framework for predicting suction nozzle health cycles

DT:
Radium engraved barcode allows the suction nozzle to have a separate ID, and IoT technology connected to the machine allows for the real-time collection of data such as the amount of suction, waste rate, yield of parts, and so on
AT:
Establishes the health cycle model of the filter element of the suction nozzle, allows for the prediction of the lifecycle of the suction nozzle, and completes the automatic identification of how dirty the filter element is
OT:
Operators can further develop automatic nozzle replacement mechanisms based on the data analysis results and accurately evaluate the amount of raw materials and effective production scheduling
PT:
Establishment of a suction nozzle health cycle early warning platform, the model part belongs to the platform, and through App notification of anomalies, real-time monitoring of suction nozzle usage status is realized

4.2.1.2 Case Study 2: Machine Tool Remaining Useful Life Prediction

Another case study is predictive maintenance for machine tools. The core component of mechanical processing is the tool used to cut the work piece, which has the most direct impact on the quality of the work piece. In the process of machining, the tool will wear gradually depending on the quantity of work performed, which will result in the decline of processing efficiency and quality. After a certain amount of wear on the part, it should be replaced. If there is no replacement, serious damage will be done to the machine, resulting in greater losses.

Foxconn has more than 175,000 different types of precision CNC equipment, which requires a lot of manual work to monitor and manage the processing conditions and the wear on the cutting tools, as well as to determine when the tool should be replaced, which is based on the experience of the workers. Tool management has many uncertainties, such as processing parameters, work materials, and other factors which affects the tool wear speed. Therefore, it is difficult to judge the tool lifecycle simply by using empirical rules.

Tools of different specifications will be widely used in the process of machine processing and die manufacturing of mobile phone structural parts. For field operators, the residual life and wear of the tool is invisible information. In the past, intelligence could determine cutting time through experience, or by the experience of operators through the observation of the cutting flame and sound to determine the tool's status. With the aim of worry-free cutting tools, Foxconn's core research and development (R&D) project, and an important subject in the field of global mechanical processing, has been to make the state of cutting tools real-time and accurately evaluated and to improve the stability and life of cutting tools. In the process of achieving the goal of worry-free tools, Foxconn adopted the problem decomposition method and solution path as shown in Fig. 4.6.

4.2 What Will Become the "Killer Applications" ...

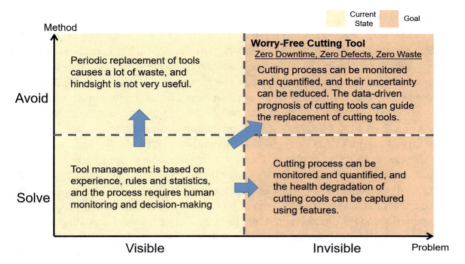

Fig. 4.6 Technological implementation path of worry-free tools

There are many factors affecting tool life and stability: tool material, structure, coating, and performance, as well as the product's material and structure, processing cooling effects, site environment, etc. Moreover, the machining process is complex and difficult to control, which affects the tool's lifecycle or variable irregularities within a tool's lifecycle. The pain and bottleneck for traditional tool life management is due to the inability to accurately predict the normal wear, chipping, breaking, and other conditions during the process of tool processing (see Fig. 4.7). Tool life management is carried out through the experience of the processors, the processing time, or the cutting length of the tool. However, replacing a tool prematurely will result in increased costs, while delayed tool replacement may result in abnormal quality, and may even cause significant damage to machine tools.

Enterprises need a certain amount of manpower to monitor and detect product quality, as well as to bear the loss associated with abnormal products. Therefore, a mechanism for tool life monitoring and prediction is needed to improve the efficiency and quality of cutting. As the direct executor of the cutting process, the tool will

Fig. 4.7 Tool cutting wear process (from left to right)

inevitably wear, collapse, and break during the cutting process. The change of the tool's state will directly lead to the increase of cutting force, cutting temperature, surface roughness of the product, product size overshoot, chip discoloration and cutting vibration, etc. In the traditional mechanical processing industry, the health status of cutting tools is judged by chip color, processing time, noise, and off-line measurement, which all require a lot of manpower cost or testing time. In addition, the traditional mathematical analysis model cannot meet the actual needs under the condition of high frequency data sources and various factors. Therefore, it is necessary to use Industrial AI technology to analyze and model the huge amount of data in order to solve the problem of tool life monitoring and prediction.

In order to monitor and predict tool life effectively, first, regarding the data technology, edge intelligent hardware is deployed on the target machine tool, and the collected raw data is processed by signal processing and feature extraction. It is then transmitted to the Fii Cloud computing platform via high power computing. Relying on edge computing technology, more than 400 key features can be extracted to characterize the tool wear state, which reduces the volume of data transmitted by nearly 100 times, which effectively reduces the burden of data transmission and computing power and reduces the investment cost of the infrastructure, such as communications equipment. After data acquisition, data availability is evaluated before the model is established. Mathematical models and parameter training are established by using effective information to avoid poor quality data affecting the prediction model.

In this case, we collect high-frequency data from sensors and controllers and low-frequency data from programmable logic controllers (PLC), including vibration signal, current signal, single processing section, and processing time. After data preprocessing, segmentation, and feature extraction, we extract different kinds of time-domain and frequency-domain feature sets and use different types of automatic feature selection methods. The feature selection method is used to select different features, and the tool wear estimation model can be established. Based on the estimation results of the tool wear, the tool's remaining useful life prediction model is then established.

Finally, the model is deployed on the service platform to develop the customized interface for the upper application, providing the interface for the designated tool to transmit real-time data, and allowing for the monitoring and prediction function of the tool's remaining useful life. The technical framework of this example is shown in Fig. 4.8.

Tool life monitoring and predictive maintenance systems (see Fig. 4.9 and Table 4.2) can minimize maintenance costs and optimize product quality. According to the preliminary statistics, the system can reduce 60% of unexpected downtime, 50% of the manpower required to inspect and monitor the status of the machine, and the quality defect rate is reduced from 6 to 3, which can save 16% of the tool cost every year.

4.2 What Will Become the "Killer Applications" ...   73

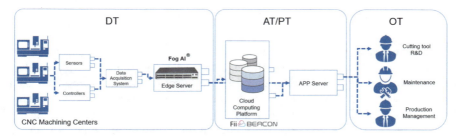

Fig. 4.8 Technical framework of tool life monitoring and prediction

Fig. 4.9 Tool life monitoring and prediction maintenance system

Table 4.2 Technical framework for tool life monitoring and prediction

DT: The edge intelligent hardware is deployed to collect the low-frequency data from the machine tool's PLC by bus communication, and the high-frequency data from the spindle current and vibration are collected by external sensors
AT: • Evaluate the availability of data to provide performance guarantee for subsequent modeling • Establish tool wear model to provide tool wear estimation • Establish tool remaining useful life prediction model based on wear estimation
OT: Online monitoring and predictive maintenance system can help staff monitor tool status, effectively use the platform to manage tool ordering and production planning, and improve staff work efficiency
PT: An online monitoring and predictive maintenance system for cutting tools is built, and the model part belongs to the platform to allow for real-time monitoring of tool decay

4.2.2 Virtual Metrology and Process Quality Control

As communication technology becomes more mature and extensive, factories are faced with more sensor data, controller data, quality measurement results, various manufacturing systems, and maintenance history. How to process this data to find out the key impact parameters for product quality is undoubtedly the core competitiveness of the manufacturing industry for the global Industry 4.0 revolution.

First, we take as an example the application of a multi-stage process scenario in quality prediction, which can benefit the manufacturer by inferring the results of quality detection in advance, and further improve the current process by tapping into the underlying causes. In the future, it can be extended to the upstream and downstream manufacturing value chain of suppliers and customers.

In recent years, in response to the development of the Industry 4.0 concept as well as the maturation of IoT technology, enterprises have been able to collect relevant data from the production line during operation. With the maturation of production monitoring and industrial data analysis technology, enterprises have begun to think about how to reduce in-plant quality measurement costs in addition to internal failure costs through quality control (QC) technology. The process quality correlation has developed into an important research direction in the field of quality management. It mainly uses data exploration technology to derive knowledge that is both meaningful and not easy to observe from the huge process data. With increasing amounts of information, data exploration methods are increasingly used in various industries to handle data analysis.

In the manufacturing industry, product quality is an important indicator for evaluating the production capacity of an enterprise. The quality results of production are influenced by many factors and are usually related to each other. Generally speaking, engineers often find it difficult to quickly and efficiently find out the actual reasons for abnormalities by relying on their own professional knowledge or experience. Although highly experienced engineers can estimate the cause of a problem based on experience, they still cannot clearly provide the specific cause, but can only speculate, which contains a higher degree of non-compliance, less stability, and very low accuracy. Especially in the Multistage Manufacturing System (MMS) (Fig. 4.10), the problems faced have more complex variables. The interaction and accumulation of many factors will affect the final quality. It is difficult to restore production to the same level as prior to a system failure, let alone be able to trace the root of the problem and how to ameliorate it.

In order to solve these problems, machine learning technology and statistical methods are widely used in multi-stage process system data analysis to help understand and provide direction for improving process quality. Finding quality-related rules with the help of relevant algorithmic tools can help clarify the relationship between quality and variables, help improve the process, and improve the quality yield. Classification and association rule mining are two kinds of commonly used algorithm tools. Classifiers include naive Bayes, principle component analysis (PCA), support

4.2 What Will Become the "Killer Applications" ...

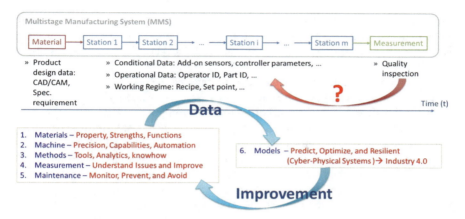

Fig. 4.10 Quality prediction and improving continuous process stations

vector machine (SVM), k-nearest neighbors (KNN), artificial neural network (ANN), etc.

For most analysis processes, the data from the final station results are directly used for analysis and the qualitative influence between machines is not considered. In this study, the influential relationship between the machines in the MMS is considered, and the quality rules are extracted by the decision tree. Then, they are constructed. By establishing a quality prediction model, we will find the data on the defective product prediction results based on the prediction model, and further analyze the causes of defective results through an a priori algorithm.

In the manufacturing industry, product quality is an important indicator when evaluating the production capacity of an enterprise. There are many factors that affect the quality yield and they are usually related to each other. Finding out the quality-related rules through the relevant algorithm tools can help to clarify the relationship between quality and variables, classifier, and rule exploration. The algorithm can extract rules from the process quality data.

A classifier is a model of data classification based on the information provided and its characteristics of class. The model can provide information about the characteristics of various classes of attribute data and can be used to predict the class of new input data. A decision tree is the most common method for classifiers.

Decision trees refer to a tree-like structure in which the intermediate node (non-leaf nodes) of the tree represents the test conditions, while the branches of the tree represent the test results. Leaf nodes in a tree represent the classification markers obtained after data classification and represent the results of classification. Since the 1960s, many scholars have used tree structures to analyze data, including ID3, C4.5, and CART. ID3 and C4.5 decision trees are only applicable to category variables. If there are continuity attributes, advanced data conversion is necessary. On the other hand, CART decision trees [4] have no such restrictions. Decision trees are a widely used tool for classification and prediction. Also, since decision trees are a tree-based method, they have the advantage of allowing for greater understanding. Generally speaking, the accuracy of a decision tree depends on the number of data sources. If the decision tree is constructed based on huge data sets, the predicted results are usually in line with expectations.

Association rule mining is a method of finding out the relationship between variables in data sets. Initially, association rules were proposed by Agrawal et al. in 1993 [5]. He determined them according to the strength of variable relation rules in data. Then, Agrawal and Srikant proposed the a priori algorithm in 1994 [6]. This method found frequently occurring items and actively established relationships between them. Then, through the process of elimination, the algorithm would eliminate less frequently occurring item sets to arrive at a rule.

To verify the feasibility, we use SECOM (Semiconductor Manufacturing Data Set) as the experimental data set [7]. SECOM is the manufacturing data for semiconductors. In this data set, 1567 pieces of datum are included, each of which has 590 manufacturing operation variables and corresponds to a quality result. 1463 pieces of datum are good products and 104 pieces are bad products. The data in SECOM are collected by sensors in the semiconductor manufacturing lines. The naming of manufacturing operation variables is based on the number of sensors. After pretreatment, SECOM will remove 115 useless features and retain 40 of the most important key features after feature selection. In order to simulate the data situation in MMS, the 40 key features selected will be divided into five feature groups, each group according to the recommendations in the literature. Each represents the manufacturing operation variables for the manufacturing machine.

Before the analysis, the relevant features are obtained through PCA. The features of each station are combined with those of the front station, and then PCA is used to acquire the associated features for the home station. The data set is processed sequentially until the last station is completed. Finally, 14 associated features are included in the data set and a quality result is obtained. After the transformation, the associated feature data set will be separated into training data and test data. The proportion of segmentation is 75% and 25%. After separation, 1175 units of training datum and 392 units of test datum will be obtained respectively. Because the data in SECOM are all numerical data and some algorithms can only deal with categorical data, it is necessary to discretize the data to provide the corresponding data types. The method of data discretization in this study is based on the concept of a control chart to divide the data's hierarchical regions. The data distribution will be divided into six types, including "very-low", "low", "normal-low", "normal-high", "high", "very-high", and so on. The final results show that the prediction quality of the associated rules mined by the Rough set algorithm can get the best accuracy, while the decision tree classifier can achieve the best performance with C5.0.

If we can find the relationship between hundreds of process parameters and final product quality parameters in a multi-stage production system, can we predict quality before manufacturing is completed? This can not only predict the risk of quality deviation, but also save a lot of time for measurement, further improve manufacturing efficiency, and reduce waste of defective products. In fact, similar attempts have been carried out for a long time in semiconductor manufacturing. The initial exploration in semiconductor manufacturing was mainly due to the higher economic losses caused by defective products, and more than 30% of the time spent in the process of semiconductor manufacturing is used for quality measurement in the intermediate process.

In the semiconductor manufacturing process, wafers often need to undergo hundreds of independent processes before they become finished products.

Because of the complexity involved in the manufacturing steps, which inevitably leads to the accumulation of manufacturing errors, it is particularly important to detect the key quality indicators to measure the processing quality in time after some important processes. When a batch of wafers is manufactured by certain processing equipment, a small part of the sample will be sent to be inspected for quality according to protocol. However, one of the main assumptions of this sampling method is that the processing errors between different samples are statistically independent and identically distributed—that is, the processing of each product is exactly the same. Unfortunately, due to so many variables to be controlled for in the semiconductor wafer fabrication process, sometimes this gives rise to "abnormalities", so this sampling method may not fully reflect the quality of the corresponding batches.

In the context of the ever-increasing technological requirements in the semiconductor industry, how to continuously reduce costs and improve the manufacturing yield is a major concern. This section takes chemical mechanical planarization (CMP) as an example to show how Industrial Intelligence can improve semiconductor manufacturing productivity in the large data environment generated by the semiconductor manufacturing process.

4.2.2.1 Case Study 3: Virtual Metrology for Semiconductor Chemical Mechanical Planarization

The purpose of the CMP process is to grind wafers, flatten their surfaces, and prepare for the subsequent lithography process. The basic principle of the CMP process is the combination of chemical corrosion and mechanical abrasive removal to achieve global flattening. One of the key indicators of its quality is the mean removal rate (MRR).

If we further analyze the process of semiconductor manufacturing and quality control, it is not difficult to find room for improvement. Existing quality monitoring processes are shown in Fig. 4.11. During wafer processing, key variables are monitored by process quality control methods, and the wafers produced are also sampled.

This quality control method can be improved in three ways: first, the sampling method cannot guarantee the quality of all wafers in a batch. As mentioned in the previous section, the premise of spot inspection is that each wafer's fabrication process is exactly the same. However, in fact, different faults may occur during processing, which are difficult to detect by sampling. Second, this method of single variable detection is often based on threshold rules. It is difficult to detect abnormalities in

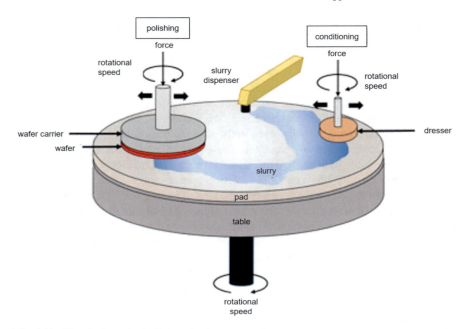

Fig. 4.11 Chemical mechanical planarization process for semiconductor wafer manufacturing [8]

the early stages of failure and it cannot predict product quality. When problems are detected, faults may have occurred in many batches, and recalls will result in delays. Third, using the sampling method of measuring equipment, when each wafer is inspected, the other wafers to be processed in this batch need to wait for the end of the testing process, which directly affects the production efficiency and prolongs processing (Fig. 4.12).

In order to solve the above "pain points", Virtual Metrology (VM) technology has attracted much attention in the semiconductor industry in recent years. VM technology is not a new concept, but through the progress of sensor technology, the steady growth of big data technology, and the improvement of computing power, Industrial Intelligence technology once again can make accurate quality prediction possible. VM refers to the use of production equipment sensor data to predict key quality indicators in the production process without actual measurement.

There are many kinds of prediction methods, which are essentially regression problems. The popular methods in VM technology include the partial least square method, ANN, and non-linear regression. More modern methods, such as deep learning, are not widely used in VM. The reason is that the performance of deep learning models depends greatly on the quantity and quality of historical samples and requires high computational power. Furthermore, the interpretability of the model is relatively poor compared with traditional methods. In industrial scenarios, how to integrate industry knowledge and experience, how to improve the real-time and reliability of

4.2 What Will Become the "Killer Applications" ...

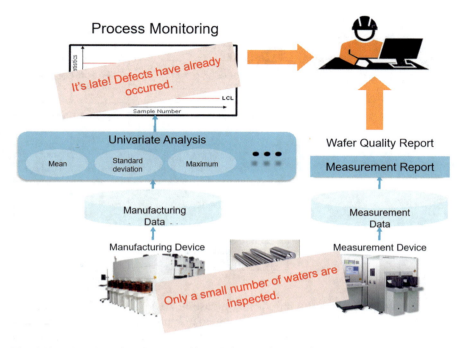

Fig. 4.12 Pain points of quality control in traditional wafer manufacturing process

the model, and how to ensure that the results of the model can be interpreted are several important issues.

In modeling practice, we find that there is no machine learning model that can perform well in all scenarios. The key to the stable output of the model is the adaptability of the modeling framework and the integration of existing domain knowledge and experience. Because the data confidentiality of the semiconductor manufacturing process is very high, in this case we chose the data from the PHM Data Challenge in 2016 for elaboration. As shown in Fig. 4.13, the key to the modeling framework adopted by the IMS Center is the integration of features and mechanisms, automatic feature selection, and integrated learning techniques.

Traditional quality inspection technology is the result of single variable statistical process control and sampling inspection. Considering the statistical distribution of a single variable, a threshold is drawn up, and the real-time value of a single variable is compared with the threshold. In order to reduce false positives, this alarm is usually tripped at the late stage of the fault. Although the traditional analytical techniques have strong mechanism correspondence, the challenge is that they can only analyze single variables without considering the influence and interaction among multiple variables. At the same time, a determination of the cause of the failure and the interpretation of the phenomena require the input from engineers and experts on the spot. It is impossible to automate and quantify the quality determination.

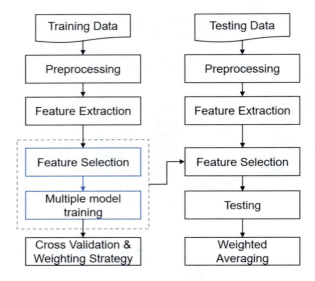

Fig. 4.13 Virtual metrology modeling framework for CMP processes [9]

This problem is solved by introducing industrial AI VM technology. First, VM technology is based on a regression algorithm in machine learning. It can use multi-dimensional process variables to predict the final key quality indicators, and then give quantitative quality prediction results.

Second, in the method shown in Fig. 4.14, feature extraction introduces mechanism-based features, including neighbor removal rate, usage features, among

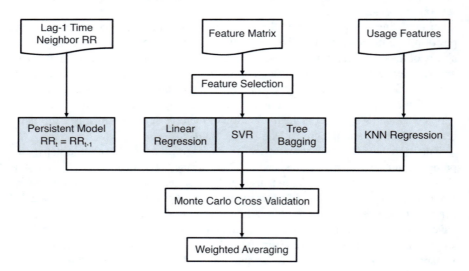

Fig. 4.14 Integrated learning framework for predicting the average removal rate of CMP process

other features. Feature extraction is one of the key steps that is able to be mechanically integrated. According to the mechanism relationship between the variables and the mechanism relationship between the variables and the average removal rate, the relevant features are extracted in a particular manner. Adding this feature to the model can provide more accurate prediction results.

In addition, T-test and out-of-bag feature selection techniques are used in the analysis process. This can dynamically select the most suitable features according to the prediction accuracy, thus ensuring that the features entered into the machine learning model are truly effective for predicting the average removal rate. Finally, the application of ensemble learning can integrate the abilities of different models and further improve the prediction accuracy. Compared with the traditional single machine learning model, the advantage of ensemble learning is that it can organically combine multiple single learning models to obtain a unified model, which is more accurate, stable, and robust.

The method introduced in this case won first place in the PHM Data Challenge in 2016. The value of Industrial Intelligence technology, including VM and predictive maintenance for the semiconductor manufacturing industry cannot be ignored. For products that are commercialized today, it is estimated that they will create more than $10 million worth of value for users, and the return on investment within a year will reach 500% ("TechEdge Prizm Overview," Applied Materials [10]). This is enough to illustrate the remarkable effect the application of Industrial Intelligence technologies such as VM will have on semiconductor manufacturing. It can not only discover the hidden dangers of quality defect in time in the manufacturing process, but also reduce the time consumed by quality inspection measurement in the intermediate process, which has great potential commercial value.

4.2.3 Energy Management and Energy Efficiency Optimization

Energy savings in factories has always been an important aspect of improving performance. Common practices include upgrading equipment, using more advanced control methods (such as frequency conversion control transformation), more rational management of energy media, and reducing the power use of equipment in the non-productive state. In almost all factories, there is plant service equipment that has an auxiliary role in production, such as equipment used in maintaining temperature and pressure, environmental protection equipment, cooling and compressed air equipment, as well as firefighting and safety monitoring equipment. This equipment is vital for the stable operation of the plant. At the same time, the plant service equipment is also a major power consumer for the factory. There are many common needs and technical means for the optimization of energy consumption for the plant service facilities. Therefore, in this chapter, I will give an example of how Industrial Intelligence can help the plant service system achieve energy savings and cost reduction.

4.2.3.1 Case Study 4: Energy Saving for Factory Service Equipment by Using Industrial AI

Factory service equipment generally refers to all facilities in the factory except for the production machines, including water and electricity supply equipment (such as ice machines, air compressors, substations, etc.), waste water treatment and emission facilities, public facilities, and so on. This case was carried out in a factory in Shenzhen which manufactures liquid-crystal display (LCD) panels. The manufacturing process of LCD panels requires a dustless environment with constant temperature, pressure, and humidity. Clean compressed air is constantly ejected from a zero-dust room, and dust and water vapor are separated through indoor and outdoor positive pressure isolation. Therefore, the demand for compressed air (Clean Dry Air (CDA)) in the plant is very large, about 221,200 cubic meters per day, converted into electricity costs, which need more than approximate 5 million dollars a year. In addition, the large-scale machine sets used in the factory need water to cool them all year round. The cooling water is provided by the ice water units in the factory system. The high-temperature circulating water flowing back from the production plant is cooled by parallel chillers and then supplied to the production plant. In addition to machine cooling, the air conditioning system of the plant also needs cooling water to provide refrigeration capacity for the air conditioning of offices and the dust-free workshop.

The power consumption of air compressors and ice machines accounts for more than 60% of the total power consumption of the factory service system. The focus of energy management is mainly on these two "big power users". The direction of energy savings mainly includes reducing energy consumption and improving energy efficiency of equipment. The air compressor and ice machine equipment in the factory yard are equivalent to the "supply end" of compressed air or refrigeration capacity, and the "use end" of compressed air or refrigeration capacity is equivalent to the "use end" of the production plant. When the supply is less than the use amount, it may affect the pass rate of production and even cause serious production accidents, so most factories will set certain amounts of compressed air for reserve usage. However, too much redundancy will lead to energy waste, so it is necessary to reduce energy consumption as far as possible but only in the context of ensuring safety in production. On the other hand, because the energy consumption of equipment in normal operation is huge and relatively stable, the energy-saving benefits brought about by improving the energy efficiency of equipment are considerable. Common factories will regularly repair and maintain such high-power equipment to ensure that the equipment operates in a healthy state with high energy efficiency.

4.2 What Will Become the "Killer Applications" ...

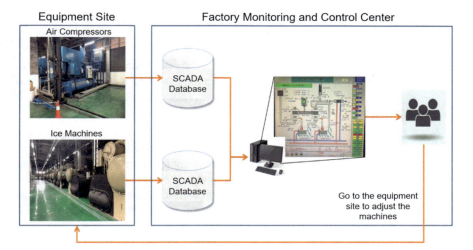

Fig. 4.15 Operation of traditional FMCS

The traditional Facility Monitoring and Control System (FMCS) (Fig. 4.15) collects real-time data for equipment through the supervisory control and data acquisition (SCADA) system and presents it on the computer screen in the central control room. It can automatically determine whether the upper and lower thresholds for the equipment are exceeded and set off an alarm. The staff on duty in the central control room will check the parameters (such as the outlet pressure for the air compressor, the opening of the intake valve, the load percentage of the ice machine, etc.) of the equipment at any time, monitor energy waste (such as the opening valve of the air compressor going below 65%) and decreases in energy efficiency (such as the ice machine overheating). According to their experience, they will be able to determine if there is a problem. When problems are found, they call the operator on site or attend to the equipment site to start, stop, or maintain the equipment.

The main problems of the above-mentioned traditional FMCS are as follows:

(1) Human experience may be outdated and not optimal, and experience is difficult to pass on. For plant facilities (air compressor, ice machine), control instructions are usually issued by experienced plant engineers or formulated through control strategies. According to the service life of plant equipment and the equipment parameters at that time, depending on how much experience the engineers have with the equipment, the equipment can be safely and reliably regulated. First, these experiences are not necessarily optimal. It is possible that changes in the field situation are beyond the experience of engineers. Second, these experiences are difficult to pass on to new people. They can only be expressed by some mechanical rules, and knowledge transmission is difficult and very slow.

(2) It is difficult to adjust the response time depending on people's duty. The granularity of start-up and shut-down control of plant equipment is determined by the frequency of personnel observation. The same duty personnel cannot be

on duty 24/7 in front of the monitoring computer and observe attentively. The control will be delayed compared with the actual situation, which is wasteful and unsafe.

(3) The maintenance cycle and schedule are not flexible enough. Maintenance of plant equipment is usually carried out on a regular basis, every six months to a year for overall maintenance, quarterly for replacement of consumables. This maintenance cycle is fixed, and not the most economical.

Some factories will install a set of automatic control procedures to regulate the first two problems. Automatic control systems can be adjusted in real-time according to some set values, such as the air compressor system monitoring pipeline pressure. When the pressure is low it will start the air compressor, and if the pressure is high it will likewise reduce the air pressure. This regulation is nearly real-time and can reduce the delay to a great extent, but this control strategy is still "short-sighted", focusing only on the present, not considering future situations. Rapid and large fluctuations in demand may lead to the regulation system exceeding its capacity. The air compressor and other high-powered equipment are not allowed to start and stop frequently; starting current is three times the normal operating current, which can cause damage to equipment and the power grid. Unfortunately, there is still no established solution to the third problem in the industry.

In order to solve the problem of the traditional energy management system for factories, industrial data analysis combines the intelligent algorithm with the operation mechanism of the plant equipment, and proposes a new solution, which can respond to the demand of the plant in advance and monitor the declining trend of the energy efficiency of the equipment, so as to achieve predictive maintenance. Specifically, industrial data analysis is mainly used for theoretical modeling, prediction algorithms, optimal decision-making, and health assessment.

1. System Mechanism Modeling

The whole system for the plant services is modeled according to the operation mechanism and historical operation data of the plant facilities. The input of the model can be a set of adjustable parameters on the equipment, or it can be opened or closed at a certain point in time. The actual output is a parameter to measure the operation status of the whole plant's facilities. For example, if the outlet water temperature of the ice machine is increased by 1 degree, the load of the ice machine will decrease, and the refrigeration capacity of the whole ice machine system will decrease. The function of the model is to fit the relationship between these physical variables. This theoretical modeling is the basis for the next decision-making optimization. In fact, there is only one real physical world. We cannot assume that if we change the decision at that time, there will be a better effect, so we need to simulate it on the theoretical model. As shown in Fig. 4.16, using the same actual input, the actual output and theoretical model output are measured by the error. The difference between the optimized theoretical model output and the actual input theoretical model output is the result of decision optimization. In the first step of theoretical modeling, the modeling error should be minimized, and the improvement of the latter's optimization effect should be paid attention to in decision-making optimization.

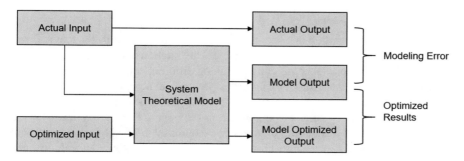

Fig. 4.16 A schematic diagram of theoretical modeling

2. Demand Forecasting Algorithms

Precise and advanced decision-making is based on the ability to accurately predict changes in future demand, which is a combination of machine learning and mechanism modeling. For example, the change in refrigeration demand at the plant end generally has two causes. The first is the change in external air temperature. It is usually hot in the summer or at noon, so at this time the demand for refrigeration is greater, while the demand for refrigeration is less great during the winter or at night. Time series prediction and long short-term memory (LSTM) models can be used for this change.

The second kind of change is caused by manufacturing processes. For example, there is a "waste solvent recovery" process on the production end. When the demand for refrigeration is very high, and an ice machine is often needed, while the production end and the factory end are split. The changes at the production end will not be notified to the factory in advance, so the refrigeration capacity is often insufficient. These changes can be predicted by data fusion, mechanism modeling, event-driven technology, etc. The main problem is to connect the data at the production end.

Looking at this from a different dimension, demand forecasting can be divided into short-term and long-term. Long-term is based on a large amount of historical data, combined with external gas environmental factors, the production scheduling plan and the daily production plan (the more products are produced, the higher the crop momentum of machine opening, the more gas consumption and refrigeration needed) for analysis in order to get a day or a shift of factory equipment scheduled. However, real plans often fail to catch up with change, so a short-term prediction model is needed to consider real-time changes. The algorithm for a short-term forecasting model is data-driven (without considering production and commissioning plans). It can avoid the planned but unrealistic situation by predicting the trend of large changes in data. The combination of long-term and short-term predictions can precisely advance decision-making.

3. Decision-Making Optimization Algorithm

With the same refrigeration or gas consumption, the number of different devices and the energy used by different devices is variable. The function of the optimization algorithm is to find the optimal configuration. Taking different equipment configurations as input and putting them into the established theoretical models, the final economic indicators are obtained as the output. Considering the constraints on demand satisfaction, the start-stop rules of equipment and the energy efficiency level for the equipment, the equipment configuration can be optimized. The algorithms used are linear programming, genetic algorithms, and particle swarm optimization algorithms, among other optimization algorithms.

The demand obtained by the prediction algorithm can be used as the condition of the optimization algorithm to optimize the start-up and shutdown decision of the equipment (see Fig. 4.17). From the long-term forecast results, we can get the scheduling situation for plant equipment the next day or shift and post it as a bulletin in the plant central control room to guide the start-stop scheduling of equipment. Short-term forecasting results can provide more accurate decision-making for immediate start-up and shutdown, which can be used to recommend the regulation of plant equipment. This proposal has considered the limitation of start-up and shutdown rules in the algorithm and can be directly implemented.

4. Health Assessment Algorithms

Combined with the PHM method, the energy efficiency of ice machines and air compressors can be monitored. Based on the analysis of current and vibration signals, the index of energy efficiency and health status can be found. When the energy efficiency decreases due to equipment decline, a comparison is made between the

Fig. 4.17 Suggestions optimizing air compressor startup and shutdown based on compressed air demand prediction

4.2 What Will Become the "Killer Applications" …87

Fig. 4.18 Plant equipment health management and predictive maintenance process

cost of maintenance and replacement of consumables to find the optimal time point for maintenance and remind maintenance personnel of the relevant losses (Fig. 4.18).

The overall functional architecture can be summarized in Fig. 4.19. First, in the data layer (also known as the edge layer), the operation data from various devices, the data from sensors, and the data for production schedules can be accessed, the data collected, processed, and stored through the infrastructure as a service (IaaS) layer for use by intelligent algorithms. In the platform layer, big data processing, analysis based intelligent algorithms, and machine learning will be used to develop a series of models needed by plant operation systems, such as energy efficiency evaluation, demand forecasting, cost calculation, etc. Finally, it can realize the application of

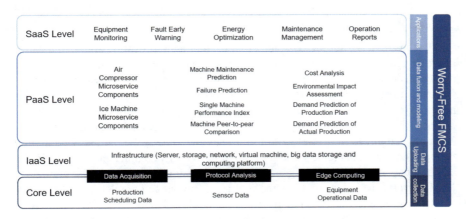

Fig. 4.19 Technical architecture for integrating the functions of intelligent plant energy management system

various industrial scenarios, such as optimization scheduling, start-stop suggestions, predictive maintenance, and so on.

The effects of the Smart Energy Management System in the implementation of this case include:

(1) Changing human observation and experience-based regulation into the regulation based on system intelligence suggestions;
(2) Changing the lagging stress regulation into predictive regulation;
(3) Changing regular maintenance of equipment into real-time monitoring of equipment status and predictive maintenance alarm.

By calculating the historical data for the whole year of 2018 and comparing the cost before and after optimization, the economic benefits are calculated as follows: 10 air compressors with 1500 horsepower can save more than $300,000 a year, and 12 ice compressors with 136 cold tons can save more than $70,000 a year.

In addition to power consumption optimization, cost optimization can also be carried out on the power supply side. This requires that the factory establish a set of multi-energy complementary systems covering renewable energy sources such as the power grid, energy storage facilities, solar energy and wind energy. This will allow for the integration and intelligent operation of different energy supplies.

Multi-energy source complementarity is a kind of energy-using mode which can alleviate the contradiction between supply and demand from multiple energy sources, protect, and rationally utilize natural resources and obtain better environmental benefits according to different resource conditions and energy-using objects. Multipurpose complementarity has the following two characteristics:

(1) Includes a variety of energy forms, constituting a rich functionally structured system.
(2) The multi-energy sources complement each other and use a step-by-step approach to reach the effect of $1 + 1 > 2$ so as to improve the comprehensive utilization efficiency of energy systems and alleviate the contradiction between energy supply and demand.

The multi-energy source complementarity intelligent operation and maintenance platform completes the collection and storage of the data from the wind turbines, photovoltaic equipment, energy storage equipment and the power grid price. At the same time, the historical data and real-time data are used to dig deeply, and the machine learning algorithm is used to establish the prediction model for the power demand, the prediction model of power generation capacity on the power supply side, and intellectualization of the control optimization strategy, health assessment, and fault prediction model for key equipment of multi-energy complementarity platforms are established to realize intelligent operation and maintenance and decision support.

The overall technical route of the multi-energy source complementarity intelligent operation and maintenance platform is based on Hadoop big data technology architecture. It is constructed of a technical platform framework for enterprise-level industrial large data storage and data visualization applications. It is a set of data collection, data extraction/processing, large data storage, large data analysis, large

4.2 What Will Become the "Killer Applications" …

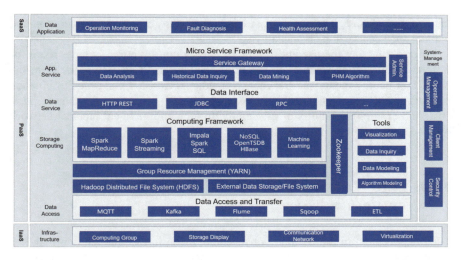

Fig. 4.20 Technical architecture of multi-source energy complementarity intelligent operation and maintenance platform

data mining modeling, operation, and maintenance monitoring. Integrated large data is integrated with the application platform to meet the application needs of remote condition monitoring, fault prediction, and health management of power generation equipment.

The overall technical architecture of the multi-energy complementarity intelligent operation and maintenance platform is shown in Fig. 4.20.

The multi-source energy complementarity intelligent operation and maintenance platform applies key technologies such as big data, cloud computing, IoT, AI, etc. It provides a variety of storage schemes and data algorithms, supports the collection, storage, analysis and mining of structured data, semi-structured data and unstructured mass data, and provides a variety of open interfaces. It also supports the redevelopment of the micro-service mode. At the same time, the platform provides a visualized data display, modeling and analysis, data management, system management and other tools, which reduces difficulty for data analysts, system administrators and end users.

The multi-source energy complementarity intelligent operation and maintenance platform can provide evaluation and report forms for the overall operation efficiency of all the equipment in the multi-source energy complementarity system (wind turbines, energy storage equipment, photovoltaic equipment, and seawater desalination equipment). The overall operational efficiency of the equipment is evaluated, and the operation result report is given. Also, it can establish a complete fault early warning system and fault diagnosis system to enable intelligent operation and maintenance as well as decision support.

According to our practical experience and data statistics, the application of intelligent operation and maintenance platforms in the field can improve the field response to faults, processing and the spare parts co-ordination process, and deal with hidden faults ahead of time. The accuracy rate is higher than 85%, and the operation and maintenance costs can be reduced by up to 20%.

After optimizing the use and supply of energy consumption, is there room for further energy savings and cost reduction? We can also find new space from the balance of energy media and the energy pipeline network. Taking the gas pipeline network made of iron and steel as an example, how to reduce energy consumption costs and ensure the stability of gas supply through an intelligent pipeline network balance system will be introduced next.

4.2.4 Defect Detection and Material Sorting Based on Machine Vision

The breakthrough in the accuracy of machine vision in face recognition makes it a killer application for Industrial AI. Many scenarios emerge around face recognition technology, including facial recognition payment methods, intelligent security, social networking, public area security, and other applications, which can greatly improve and change people's lives. The application of machine vision technology in industry is also developing rapidly. Compared with facial recognition, the accuracy of machine vision recognition in industry is higher, the speed is faster, and the detail recognition ability is also higher. Taking the appearance defect inspection of the production of the iPhone as an example, it needs three manual inspections under six directions of strong light. The inspectors are trained to recognize scratches with a depth of more than 2 μm. It is still difficult to achieve fine recognition by machine vision at present. The objects used in industrial environments are more diverse and complex, and they are often applied only a few times in the process of a product being manufactured. This makes model training more difficult, and the algorithm needs to achieve faster convergence and continuous stability of accuracy in fewer labeled samples. This will be more difficult for small-batch, multi-batch, and highly customized industrial products. Therefore, machine vision is often used to detect quality defects in industrial scenarios, such as 3C (Computer, Communication and Consumer) electronic manufacturing, automobile manufacturing, and other large-scale production scenarios.

In the domain of automobile manufacturing, most of the automobile body panels and components are sheet stamping parts. Stamping is the first of four processes of punching, welding, painting, and assembling. The level of stamping technology and stamping quality are very important for automobile manufacturers.

4.2 What Will Become the "Killer Applications" ...

4.2.4.1 Case Study 5: Intelligent Quality Management System for Car Body Stamping Process

There are three stamping production lines in the stamping workshop of a well-known domestic automobile manufacturing enterprise, which mainly produces passenger car body panels with large profile sizes and curved surfaces, such as sidewalls, fenders, doors, and engine covers. In the stamping production process, there are many factors affecting the local cracking of the side wall during the drawing process, such as the performance of equipment, the state of the die, the performance of the sheet metal, the temperature of the workshop, and so on. Some new models for the side wall are prone to local cracking in the drawing process, which requires repeated parameter adjustment and trial-production, and has a certain level blindness. This leads to high costs and low efficiency.

Based on the core trouble points involved in stamping on an automobile body manufacturing line, through extensive investigation and scheme optimization, an intelligent platform for determining and optimizing the quality of appearance for stamping products based on machine vision has been developed (Fig. 4.21). Through the construction of a large data management and calculation platform, all the equipment, dies, materials and manufacturing of a stamping workshop in automobile manufacturing enterprises have been realized. The integration, storage, and unified management and control of processed data and quality inspection data have merged together in one place to solve problems from large differences in data forms, scattered data sources, real-time signal data of equipment, and unstructured image data. The system provides data standards and safety management systems, while fully maintaining enterprise data assets. With the help of data mining based on machine learning and intelligent detection technology based on machine vision, companies can better make predictions about the punch cracking on opposite sides and the intelligent recognition of product surface defects.

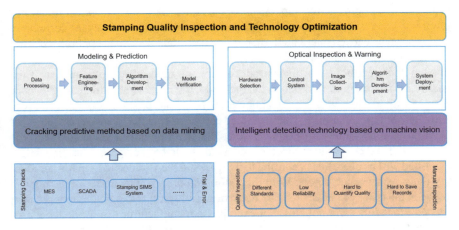

Fig. 4.21 Intelligent quality management scheme for automobile stamping production line

Based on machine vision and current conditions on production lines, an image acquisition system can be designed. Through image sample acquisition and machine learning intelligent construction, an intelligent recognition model with fast recognition speed and high recognition accuracy can be established. The model can quickly detect defects in newly produced punched-out parts and then save all the inspected images and all the processed data from production to the big data platform. Through the correlation among quality inspection data, production process and product design parameters, and with the help of large data analysis technology, a closed-loop connection for analysis and management of stamping product quality problems can be formed, allowing for the precise control and optimization of stamping product quality.

After the real-time feedback of stamping product quality is obtained, the intelligent prediction model of the stamping process can be established via a data mining machine learning algorithm based on stamping equipment processing parameters, sheet metal parameters, die performance parameters, and maintenance records. Through sample accumulation and model training optimization, the cracking risk for stamping parts can be accurately predicted. Finally, the correlation among the factors affecting the manufacturing process can be determined, and the combined control strategy for production process parameters can be formulated to support the process optimization and quality control of stamping manufacturing process.

Through the formal deployment and implementation of this platform in a production base for an automobile enterprise, the cracking risk of stamping parts for new models can be predicted, which will greatly improve the design efficiency of stamping parts processing parameters for new models. It can also reduce the number of trial production by about 70%, and reduce the trial production cost by more than $3 million per year. Through machine vision technology, rapid and intelligent detection of surface quality defects of stamping parts greatly improves the stability and reliability of production line detection, reduces the labor intensity for quality inspection workers, and can save more than $150,000 annually on the three production lines of the base stamping workshop. At the same time, the quality inspection data for stamping products are stored effectively on the platform, and relevant correlation analysis can be carried out, which can provide important decision support for quality closed-loop analysis and traceability.

Manufacturers hope that their products have no defects or problems, but if the quality inspection is carried out on a large scale, it will inevitably increase the labor cost. Machine vision is an ideal technology to help enterprises automate such problems. The Shelton Vision [11] has a surface detection system called WebSpector that identifies defects, stores images, and stores collateral metadata associated with the images (Fig. 4.22). As the product is processed along the production line, any defect identified will be classified according to its type, and then graded to help further identify the severity of the defect. With this information, manufacturers can distinguish between the types of defects that are occurring, and it can also help them implement needed procedures and policies. For example, a manufacturer can introduce a process that, when X number of Y-type defects occur, should stop the production line and adjust a parameter on the equipment or perform equipment maintenance. This

4.2 What Will Become the "Killer Applications" ...

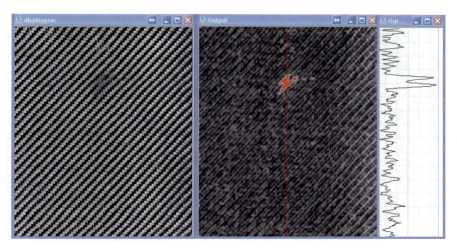

Fig. 4.22 Machine vision quality monitoring of textiles [11]

software technology, coupled with the most advanced high-speed cameras, makes system checks more than 10 times faster than manual checks. This also increases throughput, improves textile manufacturers' productivity by up to 50%, and saves millions of dollars per year from customer quality claims.

The application of AI and machine vision is not only limited to the quality monitoring of parts or products in production lines, but also can be used to help improve the health and safety of personnel within a manufacturing system. For example, Nvidia, a high-performance computer graphics technology company, has established a relevant partnership with Komatsu. Komatsu, a British-based company, is a pioneer in mining and construction equipment manufacturing. The partnership between the two companies integrates Nvidia's Jetson AI platform with Komatsu's drilling and mining equipment. With the combination of real-time camera and video analysis, the company plans to track human movement and, using deep learning, predict the movement of equipment to help avoid dangerous situations on the construction site. According to the US Bureau of Labor Statistics, about 150,000 casualties caused by accidents at construction sites are related to vehicles and machinery every year. AI and machine vision can reduce these injuries by increasing the omni-directional vision of heavy equipment. AI based on deep learning should be able to track people and predict the movement of equipment in order to avoid dangerous interactions and collisions.

Another reason for improving the intellectualization and field management of construction machinery and equipment is to improve operational efficiency. Studies have shown that up to 50% of machinery is idle in some construction sites, mainly because task scheduling and coordination of large machinery is a complex task. AI can use machine sensor data and on-site photogrammetry, including unmanned aerial vehicle (UAV) videos, to draw on-site maps in 3D, track progress, and compare their consistency with the plan. The cloud-based AI algorithm can process these visual

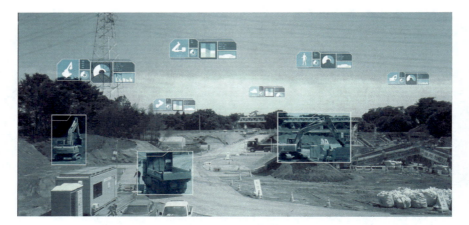

Fig. 4.23 Using machine vision to monitor construction sites and equipment in cooperation between Nvidia and Komatsu [12]

data, and create a schedule analysis model to optimize the real-time scheduling instructions for construction machinery.

There may be multiple Jetson platforms on each piece of heavy equipment. In addition to the omni-directional body camera view, Komatsu will also install a three-dimensional camera in the Jetson cab to help real-time assessment of site conditions, and rely on AI to provide additional warnings and instructions to operators (Fig. 4.23).

4.2.5 Scheduling Optimization of Production and Maintenance Plans

In industrial scenarios, both the production in a factory and the operation of a fleet attach great importance to the coordination of process and execution. In the Lean Manufacturing System proposed by Toyota, the process optimization is given important significance, which brings a new manufacturing mode, rhythmic manufacturing, to the fine design and precise execution of process. Such a manufacturing mode not only occurs in a factory, but also requires the coordination between the supply chain, logistics, and sales system. So, when we refer to the Toyota manufacturing system, we often refer not only to a Toyota factory or a production line, but also to the huge system composed of many supply chain factories which forms the pyramidal industrial chain structure of the Japanese manufacturing industry.

The application of industrial intelligent systems in intelligent production scheduling and decision-making optimization is also based on a large amount of data collection and management. By accumulating a large amount of data as well as insights into the process based on these data, we better formulate laws on how things work. Then, based on these laws, we can transform the decision-making problems into

4.2 What Will Become the "Killer Applications" ...

mathematical models, quantify the critical goal, and then define the limitations and rules for this decision. Under this framework, the best solutions that exist within countless possibilities can be found.

Traditional production planning mainly relies on the experience of the old planners. The content for production planning for traditional enterprises is often relatively singular in nature, relatively fixed in order delivery date, and does not change frequently. However, with the continuous expansion of the production demand on large-scale manufacturing enterprises and the double constraints of material and production capacity, it is increasingly difficult for planners to smoothly discharge a production plan without adjustment, which often requires the cooperation of multiple departments at the same time, and may also require constant communication with the outside world and various workers as well as repeated coordination for capacity allocation in the production execution department of the plant. Among them, the human resources required is considerable, and much more than the human resource cost is the business loss caused by a production plan that needs to be adjusted.

As an example, for manufacturing, which follows a fixed order, especially for automotive assembly lines, strict requirements in regard to execution and near-zero intermediate inventory can lead to losses if the equipment in any intermediate process is shut down for more than one minute. For example, when production is in progress, closing a machine to maintain the equipment may result in blockage of material for the upstream machine or insufficient material supply for the downstream machine. For a typical automotive production line, a one-minute shutdown can cause about $20,000 in economic losses. Surveys show that in today's American manufacturing industry, one third of the system maintenance costs are wasted due to inefficient resource utilization.

Production planning software came into being out of necessity, but at present, there are two problems in traditional planning software.

First, as the complexity of decision-making has increased to a considerable level, the issues that need to be considered are also constantly evolving. The traditional software processing scenarios often focus on the production situation in a single factory. The algorithm is usually divided into two steps: first, to determine the material, the first time to solve the problem in the case of unlimited capacity, and then to determine the production plan by tightening production capacity. Those orders that cannot be produced due to capacity constraints will be delayed. These solutions are essentially unable to take all the complex business conditions into account and add them to the model.

Secondly, because decision-making involves an exponential number of variables, the "possibilities" can often reach into the billions, so an efficient optimization algorithm is needed to ensure real-time decision-making. Some software uses a heuristic algorithm to seek a high-speed solution. When the scene is not complex, it can be adjusted and supplemented from the experience of planners. However, in light of the orders of magnitude involved with the current industrial landscape, it is impossible to solve the problem without a professional problem-solving device.

Operational research can provide solutions for these two problems, and the production scheduling problem is one of the classical problems in the field of operational research.

As production systems become more and more complex, the number of tasks and participants in the production process increases, the variety of products becomes more diverse, and the degree of flexible customization for products increases, thus many enterprises need to consider more scientific approaches when making production plans. In particular, many traditional large-scale production industries are changing to pull-type production and flexible manufacturing. Small-batch and multi-batch orders are becoming the norm. Last minute changes to orders and additional requirements occur often, which makes production planning and collaborative scheduling more difficult. If the problem is extended from a single factory to multiple factories, so that production orders flow between multiple factories and meet the needs of multiple factories with the same supply chain system, the difficulty of the problem is increased by one order of magnitude. Therefore, intelligent scheduling for many factories for small-batch and multi-batch products has become the domain of Industrial AI.

4.2.5.1 Case Study 6: Intelligent Multi-plant Scheduling Solutions

Intelligent multi-plant scheduling solutions aim at: (1) limited raw materials and fixed supply for a certain period of time in the future, (2) certain production content limitations and a fixed daily production capacity cap which cannot be exceeded, and (3) the relationship between parts and production which are stated in advance, and the time frame for parts, production, and inter-factory transit time is also set in advance. In order to satisfy the above three requirements, along with other minor constraints, the specific production content for each factory every day for the following month is optimized to ensure that the order demand within the current month meets the delivery requirements as much as possible.

Specifically, the multi-factory intelligent production scheduling solution (Fig. 4.24) not only has many constraints on material, production capacity, and transportation, but also has inventory, procurement, component production layout requirements, special raw material supply and demand, iteration updates for production relationships between upper and lower levels (BOM: Bill of Material), and component production batch version and order. There are many complicated and trivial production detail constraints such as single delivery priority. These constraints fully highlight the complexity of the problem, and the old manual scheduling models often neglect each other when constraints on such a scale need to be considered. In most cases, they can only find some scheduling plans that meet some of the above conditions, and often need to violate some constraints, while abandoning some of the order requirements. In some extreme cases, a factory's production plan may have to be suspended to wait for a feasible scheduling plan. It is conceivable that finding the only stable, effective, and even optimal scheduling plan from tens of millions of scheduling plans is of great importance to a large manufacturing enterprise. Behind

4.2 What Will Become the "Killer Applications" …

Fig. 4.24 Technical aspects of intelligent decision-making in production systems

this scheduling plan is a planning problem in which variables and constraints usually reach tens of millions or even billions of levels.

Intelligent scheduling for complex production systems and mass customization requires not only meeting current business needs, but also taking into account the multiple requirements that may occur in the future, such as utilizing the results and traceability analysis, fine tuning the robustness of algorithms and small-scale prediction errors, or full-chain optimization of a supply chain network with manufacturing as its end. In terms of the time dimension, and in light of rapid globalization, enterprises need to proactively explore more effective algorithms, coupled with concurrent processing and other engineering means, so as to improve the efficiency of solution matching. This will not only allow enterprises to meet current business demands, but also enable them to keep an eye on strategic technological reserves that have yet to be developed.

To solve this problem, the solutions provided by Cardinal Operations (Fig. 4.25) mainly include the following content: the establishment of a production planning model, the solution of the model, the analysis of production planning, the feedback of production planning, etc.

(1) Core Algorithm Model

We use linear programming for the modeling. Because of the huge scale of the problem, any integer variable will greatly prolong the solution time for the problem. In order to limit the solution time of the problem to one hour, we have tried a variety of modeling angles, successfully using some program modeling techniques to linearize the constraints and ensure the emergence of no integer variables. Specifically, we divide the solving process into several small linear programs to solve. When solving each small linear program sequentially, we not only help the solving process to achieve rapid convergence, but can also replace some integers (0–1 variables, i.e. whether to choose or not). Furthermore, we use beneficial data structures to achieve fast data conversion. This ensures that the model runs quickly and accurately from the output data to the output results.

(2) Analysis Module

When the scheduling plan is completed, the solution will provide a complete delivery analysis function for the delivery of the order, which will analyze the specific reasons for the delayed delivery or cancellation of the order, and generate the corresponding cause report (the specific material shortage, the specific factory capacity shortage, etc.).

(3) Feedback Module

Ultimately, when we identify the causes of delayed or cancelled order delivery due to insufficient material or capacity, the solution will make the most effective procurement recommendation based on the purchase list (materials that can be purchased and the expected arrival time of these materials). Through this procurement recommendation, enterprises can clearly see the improvement of order delivery directly brought by each purchase or several purchases, so as to measure whether they need to implement procurement recommendation.

In the solution provided to a large electronic equipment manufacturer by a data scientist from Shanshu Science and Technology, operational research not only met the current business needs of the enterprise, but also considered the various future needs of the enterprise. The production plan in this application has already fully absorbed the experienced planner's scheduling knowledge, the overall automation rate can reach 80%, reduce the total production, inventory, and transportation costs on the whole supply chain by tens of millions of levels.

If we move the application scenario from the factory to outside, or even to remote mountain areas or the sea, whether or not the maintenance schedule is reasonable is

Fig. 4.25 Interface of production scheduling optimization results

4.2 What Will Become the "Killer Applications" …

not only related to downtime losses, but also has a great impact on maintenance costs and efficiency. Maintenance scheduling optimization of field equipment is based on continuous monitoring and fault warning of this equipment.

Taking the maintenance scheduling optimization for offshore wind farms as an example, on the basis of accurate evaluation of turbine status, how to integrate status information, environmental information, maintenance resources, and other information is used to design a maintenance plan. The optimization of planning decision-making is also an important aspect of intelligent management and use. Maintenance of wind farms is very complex, especially in the construction of wind farms at sea. Maintenance requires special equipment such as ships, helicopters, marine engineering ships, and is more expensive and has a longer maintenance cycle. It is difficult to achieve maximum efficiency for wind energy utilization by manual condition monitoring and maintenance scheduling because of the harsh operation environment, the randomness of wind resources, and the fact that wind fields are located in remote areas. How to optimize the maintenance and maintenance tasks for wind farms based on the health status of wind turbines, the forecast results of wind resources, the availability of maintenance resources, and the weather conditions at sea is still a common problem in the industry. In order to solve the above problems, the IMS Center at the University of Cincinnati developed an optimization system for short and medium-term operation and maintenance scheduling for offshore wind farms (as shown in Fig. 4.26). The optimization model belongs to the category of non-linear optimization problems with constraints, and the genetic algorithm is used to solve the model.

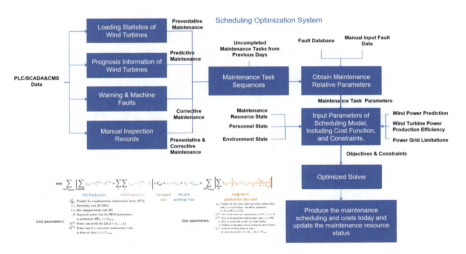

Fig. 4.26 An optimization system for short and medium-term operation and maintenance scheduling of offshore wind farms

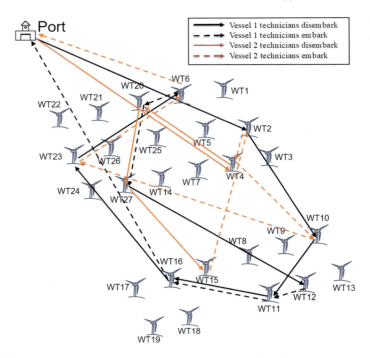

Fig. 4.27 Deduction and optimization of maintenance schedule for offshore wind fields

4.2.5.2 Case Study 7: Short and Medium-term Operation and Maintenance Scheduling of Offshore Wind Farms

Wind farm maintenance scheduling optimization (Fig. 4.27) is based on the accurate prediction of wind resources, combined with maintenance demand information, in order to choose when the wind resources are weak enough to allow for maintenance, and when wind resources are best for power generation. For each maintenance task, multiple available maintenance teams can choose to take multiple available maintenance vessels for maintenance, which increases the flexibility of system maintenance scheduling and helps to reduce costs. However, this enlarges the scope of searching and deducing feasible solutions and makes the problem more complex. A genetic algorithm is a popular heuristic algorithm for solving uncertain scheduling problems. It is strong for general applications, good computing power, implicit parallelism and global search, and has strong robustness in dealing with uncertain scheduling problems. In view of the characteristics of maintenance tasks in offshore wind farms, the deduction and decision-making environment for optimization of maintenance scheduling is established according to the idea of a genetic algorithm, taking fully into account factors such as the ships used, weather, maintenance personnel, maintenance order, health status of wind turbines, and navigation costs. This deduction environment needs to be very expansive: it needs to be able to flexibly adapt to the

4.2 What Will Become the "Killer Applications" ...

arrangement of *M* maintenance vessels and *N* maintenance teams to complete *P* different maintenance tasks. The objective of maintenance scheduling optimization is to select the most suitable maintenance vessel, maintenance team, and maintenance start time for each maintenance task, so as to minimize the losses caused by wind power loss and resource utilization during the whole maintenance process. Therefore, the maintenance scheduling optimization problem for offshore wind farms can be divided into three sub-problems: determining the start time of each maintenance task, determining the corresponding maintenance vessel for each maintenance task, and determining the corresponding maintenance team for each maintenance task.

In order to support as many ships, maintenance teams, and tasks as possible, the models in the deductive environment also need to have good computational efficiency. In the process of solving the genetic algorithm model, the two-tier nesting method is used to optimize the sequence and start time for maintenance tasks. In short, the sequence of maintenance tasks is determined first, and then the waiting time between the two tasks is determined. This reduces the complexity of scheduling problems from $N2$ to $N + N$ (N being the number of maintenance tasks).

Taking 17 maintenance tasks in an offshore wind farm composed of 24 wind turbines as an example, the cost forecasted for the execution of the optimized scheduling plan can be reduced by more than 25% compared to without the optimized maintenance plan, which is shown in Fig. 4.28. Because a meta-heuristic algorithm has strong randomness (genetic algorithm is one of meta-heuristic algorithms), it is not fully applicable to modern and standardized industrial problems requiring strong stability and traceability. Relatively speaking, this specific algorithm has the same speed as the heuristic algorithm in solving small and medium-scale problems. In terms of accuracy, it can also achieve stable cost savings of more than 25%.

For the problem of resource operation and scheduling optimization in industry, the key to finding a solution lies in whether the problem can be transformed into a certain range and the mathematical expression of the boundary problem can be

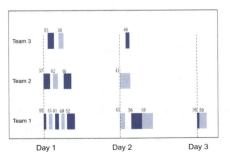
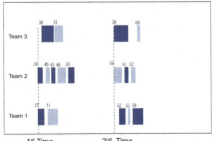

Fig. 4.28 Comparison of scheduling optimization results for 17 maintenance tasks of an offshore wind farm

established, otherwise it will fall into the predicament of excessive complexity or insolubility. This requires the precise modeling ability and rich business experience of operational research scholars.

4.3 Enabling Industrial AI Systems

4.3.1 Intelligent Monitoring and Maintenance Platform for CNC Machines

Spindle health monitoring and maintenance for the Computerized Numerical Control (CNC) machine is one good example to illustrate the application of Industrial AI at the system level. A spindle is one of the most critical components in a CNC machine tool. Unexpected spindle failure causes a huge loss of money and time. Being one of the costliest parts in the machine tool and not failing often, it is generally not kept as a spare part in a factory. This causes a huge loss in production as the replacement times for spindles can be long. Thus, it comes as a high value to spindle manufacturers to monitor spindles at the customer site and pre-ship new spindles when the current spindle is about to fail. The three qualities of self-awareness, self-configuration, and self-prediction make the spindle intelligent and highly valuable for the spindle manufacturers.

The data acquisition module in the intelligent spindle system is installed to collect high-quality vibration and monitor current data. This data is collected using the Fixed Cycle Feature Test (FCFT) technique, which is a standard cycle designed to acquire the machine 'fingerprint.' During this cycle, the machine runs at a collection of constant and variable speeds. Data collection on the machine is coordinated between the communication module in the Intelligent spindle system and the machine controller by a 'handshake.' The communication channel between the machine and data collection unit ensures the system is ready to collect data before the FCFT starts, and lets the DAQ system know when to trigger the data collection. Figure 4.29 shows the machine FCFT cycle and an example vibration data collected from the machine.

The analytic technology consists of two stages: (1) signal processing and feature extraction at the edge, and (2) data-driven modeling and cumulative learning on the server (local or cloud). The signals are processed to extract expert features based on the geometry of the bearings, motor specifications and the shaft speed. The features are stored in a database, which contains the same signature from other machines in the factory. In the server, machine-learning algorithms are utilized to establish a model that represents what a healthy machine signature looks like. Once a deviation is observed, the AI tool identifies the similarity of the observed deviation with previous observations and makes an inference about the type of fault (if any) and the time left before failure or deviation in part quality occurs.

Figure 4.30 Platform technology for an intelligent spindle. It consists of an edge-computing unit, called Cyber-box, which consists of integrated data acquisition,

4.3 Enabling Industrial AI Systems   103

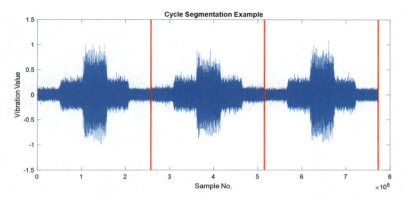

Fig. 4.29 Sample vibration data collected during a standard cycle

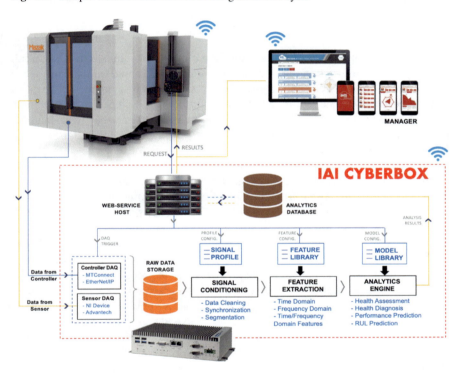

Fig. 4.30 Platform technology for intelligent spindle

Fig. 4.31 Machine health information shown in real-time in a machine controller

a communication module, and a processor for performing computations. Once a full FCFT cycle is completed, a program segments the signal based on machine speed and extracts time and frequency domain features from each segment. The features are then pushed to a database in the server for further processing. On the server, an AI program is run to extract machine health information and save that in the database along with machine information and time stamp. A web interface is designed to pull data from the database and show the machine health status and diagnosis information in real time.

The output of the health-assessment analytics makes the machine tool self-aware about its health status. Along with this, once a fault is identified by the machine tool, it can forecast the remaining useful life of the spindle, giving the machine tool self-predicting capabilities. Further, while in operation the spindle can self-configure its operating parameters to avoid faults and optimize its life. Figure 4.31 the real-time machine health information on a machine controller available for the operator to observe. This information about its health status and the remaining useful life, if a fault is detected, is reported to the machine tool manufacturer so that a replacement for the spindle arrives at the customer before the faulty spindle fails.

Figure 4.32 the 5C architecture to construct a maintenance network for the CNC machines. At the component level, the sensory data from spindle components has been converted into information, a cyber-twin of each component will be responsible for synthesizing future steps to provide self-awareness and self-prediction. At the next stage, more advanced machine data (e.g. controller parameters) would be aggregated to the components information to monitor the status and generate the cyber-twin of each particular machine. These machine twins in CPS provide the additional self-comparison capability. Further at the third stage (production system), aggregated knowledge from components and machine level information provides self-configurability and self-maintainability to the factory. This level of knowledge

4.3 Enabling Industrial AI Systems

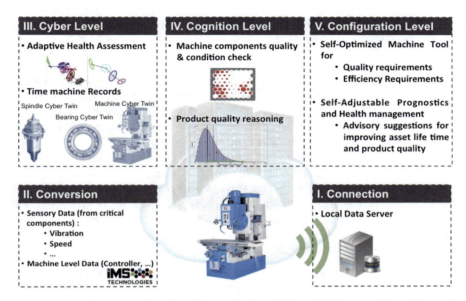

Fig. 4.32 The flow of data and information in an Industrial AI enabled factory with machine tools in production line based on 5C architecture

not only guarantees a worry-free and near-zero downtime production, but also provides optimized production planning and inventory management plans for factory management. Then the 5C architecture of CPS can be deployed in a cloud platform to serve customers all over the world (Fig. 4.33).

Fig. 4.33 A cloud-based maintenance platform for machine tools

4.3.2 Intelligent Operations and Maintenance System for Offshore Wind Farms

The wind power industry has developed rapidly over the past ten years. The increasingly fierce competition in the wind power market puts forward higher requirements for the performance of wind turbines. Wind power equipment manufacturers have made a lot of effort to reduce costs, and reducing the cost of wind turbine manufacturing is facing more and more technical challenges.

Correspondingly, at present, the operation and maintenance management of wind turbines is still carried out in a relatively laissez-faire manner, where the diagnosis and health management of turbines are still not perfect. The efficiency of operation and maintenance management needs to be greatly improved, which will provide a great space for the development of intelligent turbine systems. Based on the actual situation in wind fields, the difficulties with the development of intelligent operation and maintenance systems lie in the points listed below:

(1) Wind resources are random. The variation of wind speed, wind direction, and wind energy density are a dynamic and non-linear process. Accurate prediction of wind speed and power has always been a hot topic in academia.
(2) The locations of wind farms are mostly in remote areas or at sea. Maintenance work is complex and special equipment is often used, which leads to high maintenance costs and long maintenance cycles.
(3) Influence by the randomness of wind resources, deterioration levels, lubrication state, and wear of yaw gears of different turbine drive systems is also quite different. Traditional preventive maintenance is difficult to perform in regard to wind field operation and maintenance.

Based on the current status of wind power development and difficulties in operation and maintenance, the key functions of intelligent operation and maintenance of the whole wind farm are as follows: (1) the health management of turbine, key components, and the accuracy and transparency of recession, (2) the health trend analysis and residual life prediction of turbines and key components, (3) the implementation and evaluation of the power generation performance of turbines, (4) intelligent dispatch management of wind farms based on wind resource prediction technology, and (5) optimization of wind farm operation, maintenance strategy, and resource dispatch based on performance prediction of the wind turbine and key components.

The core of intelligent wind farm health management and intelligent maintenance systems is to evaluate the health status of key equipment and turbine subsystems, and to program intelligent scheduling and scheduling optimization for the operation and maintenance of the turbine according to the prediction of wind resources. There are many critical turbine components, and the operating conditions vary, which requires sorting out and organizing the functional levels of the system according to the realization logic. Based on the 5C technology system of CPS architecture, we designed the CPS functional architecture of intelligent wind farms as shown in Fig. 4.34.

4.3 Enabling Industrial AI Systems

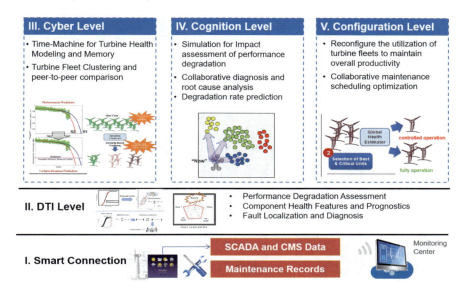

Fig. 4.34 Designs for the intelligent wind field function based on CPS 5C architecture

A wind field is a typical multi-source heterogeneous data environment. The data mainly comes from the supervisory control and data acquisition (SCADA) system and the condition monitoring system (CMS). These information sources provide environmental information, work condition information, control parameters, state parameters, and vibration signals of some key components, as well as other data. Other data sources include the power grid dispatching information, work order system, personnel management and maintenance of resource status, etc. Through the large data analysis tools provided by the Watchdog Agent toolkit from the IMS Center, the above data sources are comprehensively analyzed, and the turbine is modeled, analyzed, and visualized to form both a widely applicable turbine performance evaluation as well as a decline in performance prediction, wind farm operation, and maintenance management algorithm modules.

The whole process of data analysis and modeling is shown in Fig. 4.35. The research and development of the intelligent wind farm health management, operation, and maintenance system includes two parts: predictive analysis for turbine performance and dynamic operation and maintenance optimization of wind farms. First, through the analysis of real-time data, effective pattern recognition of current work environments and wind resources status can be carried out. Based on real-time data, effective health features are extracted to establish the health model of turbines and key components, and the current turbine degradation status is evaluated and analyzed. Based on the health assessment of wind turbines and its key components, the potential operation risks and possible failure modes of the equipment can be further judged, and the remaining effective lifespan can be predicted. This will allow for the maximization of the power generation capacity of the wind turbines, while at the

108   4 Killer Applications of Industrial AI

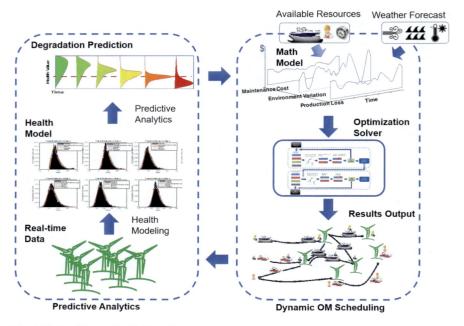

Fig. 4.35 Intelligent wind field predictive analysis and dynamic operation and maintenance process

same time reducing the downtime of the system as much as possible, so as to avoid the occurrence of major downtime faults.

The modular design of intelligent wind turbine systems makes the system more scalable. For example, deploying a cloud service system can provide more customized services for wind farms while realizing the intelligent upgrade of the wind turbines. After analyzing and processing the wind turbine implementation data, the corresponding features and models are uploaded to a cloud service platform. This can help to unify management and provide further analysis. Users can use this platform to realize remote real-time monitoring and historical performance traceability of any turbine from multiple wind farms.

Comparisons based on historical data of turbine operation and baseline model are the means to realize accurate management of the turbine state driven by data. Using different data as resources can produce different comparative dimensions and insights, including comparing changes in self-state in the time dimension and differences with other individuals in cluster dimensions.

As an example, SCADA data is rich in environmental parameters and state parameters. There is a strong spatial correlation among these parameters. This correlation is non-linear, and other parameters change when one parameter changes. The response speed is also different (for example, when the rotational speed rises, the vibration increases first, while the temperature rises slowly), so it is difficult to accurately manage with traditional physical modeling methods. When using SCADA data to carry out health management and fault prediction for turbines, the feature extraction part

4.3 Enabling Industrial AI Systems

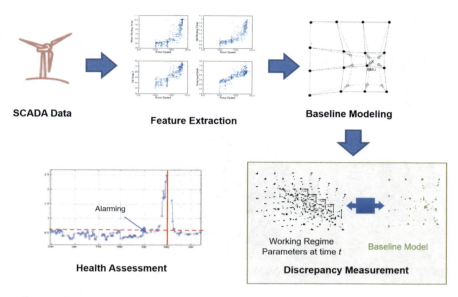

Fig. 4.36 Health assessment method based on historical turbine data

of the data-driven modeling method is improved. Instead of calculating the health value of each set of SCADA readings, the data in a fixed time window is used for modeling, and then the differences between the two models are compared.

This modeling method (Fig. 4.36) does not analyze the statistics of a single parameter or the model of the range interval, but models the relationship between parameters. When the state of components represented by one parameter in the equipment changes, this relationship will be broken, and the difference between the model and the baseline model will become larger. Through the quantitative evaluation and management of this difference, the purpose of evaluating the health status of the turbine can be achieved.

However, when the problem is too complex, and there is neither a baseline model nor state label, machine learning modeling is almost impossible to start. In the case of a wind farm encountered by IMS, the owners of the wind farm accumulated CMS vibration data from nearly 100 wind turbines for 6 months. There were 8 vibration measuring points on each fan, and three sets of vibration data were collected daily at each measuring point. There were more than 300,000 vibration documents in total. The background information related to these data is completely accurate, including the design parameters of the transmission chain, the status label of the document, and even the work condition information such as speed and power. However, these data sets hide the data of a fan with a definite fault, a serious fault of the gearbox, which occurred during this period of time. If we rely on experienced vibration analysts to analyze the frequency spectrum of the 300,000 sets of vibration data, although it can also achieve an accurate judgment, it would take an unimaginable amount of time. In this context, IMS attempts to use a cognitive modeling approach based on big

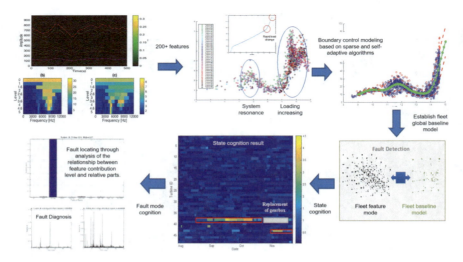

Fig. 4.37 Knowledge mining and health assessment based on cognitive model

data and cluster environment. The specific process is shown in Fig. 4.37. First, from the spectrum relation mining of all vibration signals, a frequency component of non-integer multiples is found frequently. According to experience, it is likely to be the meshing frequency component of a gearbox with two speed ratios. With one of the frequencies as a reference, all spectral components are orthogonalized. More than 200 possible health-related features were extracted. Subsequently, the spatial distribution relations of these state characteristics are depicted. Their non-linear distribution relations can be related to the dynamic characteristics and operating characteristics of the transmission chain, so they should be effective health characteristics. After further de-correlating and sparsifying of these features, the baseline state model of the whole fan cluster was established. Subsequently, the baseline for the cluster state was compared with each fan individually. After showing their difference in degree in the color temperature chart, it is clear that the difference of the No. 39 fan is obviously higher than that of other fans. Therefore, it was determined that a gear box failure had occurred, and the difference was two months after the shutdown—it was a pre-existing condition. Further contribution analysis and fault pattern recognition located the fault on the unbalanced mode of the high-speed shaft of the gearbox. After this process, new knowledge was mined from the big data environment by cognitive modeling, and unknown problems were discovered by using this knowledge.

To optimize the dynamic operation and maintenance of the wind field, it is necessary to optimize the maintenance decision of the wind field on the basis of accurate assessment of the health status of each fan, synthesizing the current fan health information, environmental information, and maintenance resource information. Prediction of wind resources is also an important basis for optimization of fan scheduling. Maintenance of wind turbines should be carried out at a time when wind resources are low, so as to reduce the loss caused by power generation and shutdown maintenance.

4.3 Enabling Industrial AI Systems

The IMS Center has developed an optimization model for short-term and medium-term operation and maintenance planning of offshore wind farms. The model is based on the actual situation of wind farms and design optimization models, and a variety of non-linear constraints to ensure that the actual operation of maintenance and dispatching sites is simulated to the greatest extent, and the optimal decision-making is given.

Aiming at the characteristics of maintenance tasks for offshore wind farms, a general optimization model for maintenance scheduling can be established, which takes fully into account factors such as ships used, weather, maintenance personnel, maintenance order, health status of wind turbines, and navigation costs. For each maintenance task, several available maintenance vessels can be selected from a number of available maintenance teams, which increases the flexibility of system maintenance scheduling and helps to reduce costs, but this also enlarges the search scope of feasible solutions, making the optimization problem more complex so that the optimization requirements can be used. Solution software such as MATLAB Optimal Toolbox and Gurobi have been difficult to use to solve this problem within a reasonable time. According to the idea of genetic algorithms, the IMS Center designed a two-layer genetic algorithm model which is suitable for solving the deduction model of the optimization of the operation and maintenance of offshore wind farms. The algorithm has strong scalability, better computing power than commercial optimization software, and is superior in dealing with the intelligent dispatching of wind farms. The interface of the intelligent wind farm operation and maintenance system developed in this paper is shown in Fig. 4.38. It includes four modules: visual output for the optimization decision, cost comparison for the optimization decision, a perspective on maintenance costs, and traceability analysis of historical costs. It can easily provide users with intelligent maintenance scheduling decision support. Taking 17 maintenance tasks for an offshore wind farm as an example, the forecast

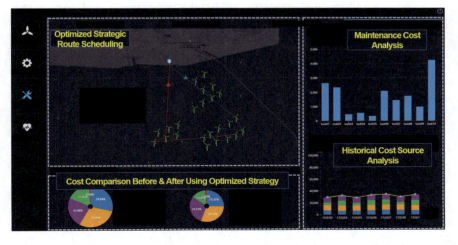

Fig. 4.38 Visual interface for intelligent wind field operation and maintenance optimization system

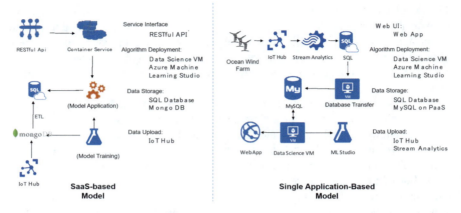

Fig. 4.39 Public cloud deployment and wind field local deployment architecture for wind field intelligent operations and maintenance system

cost for the optimized intelligent scheduling plan in the execution of deduction is reduced by more than 25% compared with that of the non-optimized maintenance scheduling plan, which significantly improves the efficiency of wind farm operation and maintenance.

The deployment mode for a wind farm intelligent operation and maintenance system can also be flexible and diverse. It can be deployed locally in the wind farm, or deployed separately on a cloud platform such as fault warning, vibration analysis, and maintenance scheduling optimization for customers to subscribe in the way of Software as a Service (SaaS). Concord New Energy is a leading wind power operation enterprise in China which cooperated with Microsoft for this project. This company and Microsoft worked together to upload wind power operation data to the Azure cloud platform in real time, and has developed the PowerPlus system to analyze and support the operation and maintenance data of wind power. Figure 4.39 illustrates the technical architecture behind different deployment modes and the differences between the cloud platform functional components used. Flexible deployment forms can meet the customized needs of different users, especially for small and medium-sized wind power operators and the marginal costs of the introduction of new technology can be significantly reduced.

4.3.3 Intelligent Rail Transit Predictive Maintenance System

As a key rotating component of the bogie system, axle box bearings are one of the core components of a high-speed railway EMU, and its health status is related to the operation safety of the whole train. We cooperated with Qingdao Sifang to carry out systematic research on fault prediction and health management (PHM) of high-speed

4.3 Enabling Industrial AI Systems

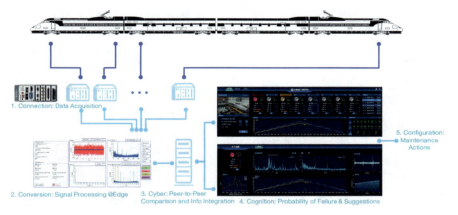

Fig. 4.40 Shaft box bearing PHM system based on CPS5C architecture

railway axle box bearings and brought about real-time monitoring of the operation status of high-speed railways, and were able to warn of and diagnose the bearing faults that can occur. This helped to avoid the hidden dangers that threaten major train operation safety and improve the operation safety of high-speed railways.

The PHM prototype for the axle box bearing is built on the basis of CPS's 5C technology architecture, and the overall structure of the technical scheme is shown in Fig. 4.40. The bottom layer is the train bearing signal acquisition and edge side calculation, corresponding to the Intelligent Sensing Layer (Connection) and the Intelligent Analysis Layer (Conversion). On the vehicle-borne edge computing platform, multi-channel vibration signals are collected uniformly, and preliminary feature extraction, screening, classification, and priority arrangement are carried out to improve the interpretability of data.

At the same time, the platform is also responsible for both communication and real-time transmission of the analysis results to the ground center. On the centralized control platform at the ground center, real-time monitoring of the health status of high-speed rail clusters is realized through the mirror model of high-speed rail bearings, corresponding to the network layer (Cyber). Finally, the results of model prediction are integrated with other information systems (such as work order, ERP, and EAM) to realize the closed-loop visualization of data presentation and operation and maintenance decision-making, corresponding to the intelligent cognitive layer (Cognition) and the intelligent decision-making and configuration layer (Configuration). In addition, it is worth mentioning that on this platform, with the analysis of model results by bearing experts and feedback from operators, new knowledge is constantly generated, and the value of the entire industrial system is constantly strengthened.

In this scenario, there are three core technologies: axle box bearing fault diagnosis and prediction, high-speed concurrent acquisition of multi-source mixed signals, and calculation of train end edge.

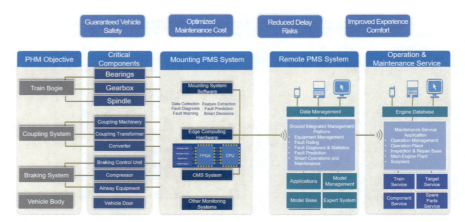

Fig. 4.41 Composition of fault prediction and health management system for high-speed axle box bearings

Axle box bearing fault diagnosis and prediction: in the modeling for the axle box bearing, the combination of mechanical models and data-driven methods is adopted. On the one hand, based on the mechanical analysis of the bearing, which is a kind of common rotating machinery, potential fault frequency band characteristics are extracted; on the other hand, based on the fault labels accumulated in ERP/MRO system, through the machine the method of machine learning trains fault detection and the classification model to quantify and diagnose bearing health risk (Fig. 4.41).

High-speed concurrent acquisition of multi-source mixed signals: in order to accurately locate and identify axle box bearing faults, it is necessary to collect comprehensive train operation and component data (sampling rate is 25.6 kHz), including vibration, speed, temperature signals, etc. Because of the large amount of data collected, in order to avoid collecting too many invalid signals, a data acquisition strategy combining event triggering and timing acquisition is adopted (i.e. acquisition of short-term high-frequency signals only when the train changes in operating conditions or enters a specific speed or reaches a specific time point). Such a data mining strategy not only guarantees the comprehensiveness of the analysis data, but also greatly reduces the pressure of edge data mining equipment and cloud data storage.

High-speed rail train edge computing (Fig. 4.42): in the high-speed rail scenario, it is unrealistic to transfer all data to the cloud server for computing and processing. Not only is the transmission bandwidth limited, it may cause data delay, but also the communication cost and storage costs are too high. In order to solve this problem, we adopted the distributed computing architecture, deployed edge computing nodes in the train, realized fast and direct feature extraction of the original data, and enhanced the analysis value of the data returned to the cloud. In this project, we extract more than ten features for each channel's vibration signal, and transform the 100-megabit raw data into Kb data pushed every 20 seconds. This greatly improved the efficiency of the system. In a distributed system, what we often say is "data transfer," but in fact

4.3 Enabling Industrial AI Systems

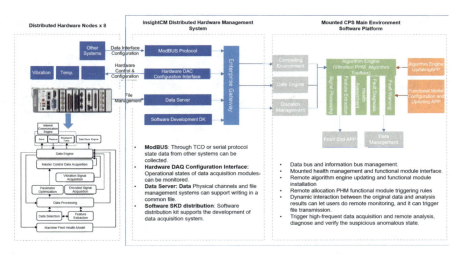

Fig. 4.42 Fault diagnosis system for vehicle axle box bearings deployable at edge end

the essence of the information being transmitted is not data, but the intrinsic value of the data. In the explosive growth of data volume, this way of thinking of value transfer can provide a wide range of possibilities for the expansion and effectiveness of the system.

The final PHM prototype system for axle box bearings includes the following two core functions: on-line monitoring of bearing health and remote configuration management of the edge end.

On-line monitoring of bearing health status: through real-time collection and analysis of bearing operation data, users can monitor the running status and health decline of the train axle box bearings remotely, and diagnose specific failure modes (inner ring, outer ring, roller, and cage) through the PHM model to provide support for operation and maintenance.

Edge-end remote configuration management: in order to meet the challenges of an edge-end data acquisition strategy and model updating after the system has been widely put into operation, the system provides a remote configuration function for the edge-end hardware. After the optimization is completed, the new model can be deployed to all train groups with one key, and users can also trigger the original data acquisition and transmission remotely for a suspected fault in the train groups.

The axle box bearing PHM system for high-speed EMU needs to provide a closed-loop to support decision-making, which means that the information it provides will generate operational insights. The optimization tool obtains the prediction information from the prediction model and peer learning model, and outputs the optimal maintenance plan, the optimal spare parts inventory quantity, and assigns the task to the appropriate personnel. Optimal algorithms and solvers are mature fields, but the mathematical modeling for maintenance tasks, resources and the output of PHM systems is still a complex project which needs to take many specifications and details into account.

Decision support tasks are usually formalized as a stochastic optimization problem. Cost function will reward the availability and function of assets, and punish the downtime of assets. The optimization problem can be solved by genetic algorithms and iso-heuristic algorithms. The challenge of solving optimization problems is often the construction of cost functions and computational efficiency. The optimization method based on simulation is high cost and time-consuming. The efficiency of the algorithm will really give users a near real-time decision-making ability (in seconds) without waiting for hours after the optimization process is completed.

With the support of this system, it is expected to bring about the transformation from after-service (problem solving) to predictive maintenance (problem avoidance) of high-speed railway axle box bearings. The dispatcher of the ground control center can monitor and analyze the fault of train bearings in real time and escort the train. Train operation and maintenance personnel can also formulate more optimized operation and maintenance strategies based on the model analysis results, which will improve the accuracy and efficiency of bearing maintenance.

After nearly two years of research and development and testing, this system has been validated on the vehicle rolling comprehensive performance test bench and test loop. The accuracy rate of bearing fault identification is over 90%, and more than 20 faults have been accumulated to further manage the uncertainty of prediction results and guide an operation and maintenance closed loop.

It can be anticipated that the first step towards intelligent health management of axle box bearings is also only the first step of intelligent operation and maintenance of the high-speed railway: after that, it will gradually expand from point to surface to realize predictive maintenance of the bogie system, traction system, braking system, body system, and door system, and finally realize the "carefree operation" of high-speed railway for passengers and provide a safer, greener, and more comfortable ride experience.

References

1. Deloitte (2018) Predictive maintenance and the smart factory. Available: https://www2.deloitte.com/content/dam/Deloitte/us/Documents/process-and-operations/us-cons-predictive-maintenance.pdf
2. Roland Berger (2017). Predictive maintenance. Available: https://www.rolandberger.com/en/Publications/Predictive-Maintenance.html
3. Lee J (2015) Industrial big data: the revolutionary transformation and value creation in INDUSTRY 4.0 Era. China Machine Press
4. Steinberg D (2009) CART: classification and regression trees. In: The top ten algorithms in data mining (pp. 193–216). Chapman and Hall/CRC
5. Agrawal R, Imieliński T, Swami A (1993, June) Mining association rules between sets of items in large databases. In: ACM Sigmod record 22(2):207–216 ACM
6. Agrawal R, Srikant R (1994, September) Fast algorithms for mining association rules. In: Proc. 20th int. conf. very large data bases, VLDB 1215:487–499

References

7. Michael M, Adrian J (2008) SECOM data set. Available: https://archive.ics.uci.edu/ml/datasets/secom
8. PHM society (2016). PHM data challenge 2016. Available: https://www.phmsociety.org/events/conference/phm/16/data-challenge
9. Di Y, Jia X, Lee J (2017) Enhanced virtual metrology on chemical mechanical planarization process using an integrated model and data-driven approach. Int J Progn Health Manag 8(2)
10. Applied Materials (2019) TechEdge prizm overview. Available: http://www.appliedmaterials.com/zh-hans/media/documents/techedge-prizm-overview
11. Shelton Vision (2019) On loom inspection – reduced waste and increased throughput. Available: https://www.sheltonvision.co.uk/case-studies/on-loom-inspection/
12. Forbes Magazine (2017) NVIDIA and komatsu partner on AI-based intelligent equipment for improved safety and efficiency. Available: https://www.forbes.com/sites/tiriasresearch/2017/12/12/nvidia-and-komatsu-partner-on-ai-based-intelligent-equipment/#2c7f40c665b0

Chapter 5
How to Establish Industrial AI Technology and Capability

5.1 Assessment of Basic Capability Maturity During Industrial Intelligence Transformation

Industrial Intelligence is rapidly changing how manufacturers operate. Enterprises preparing to adopt industrial intelligence technology should have a clear understanding of their technological abilities, as well as the ability to assess the maturity of their intelligent transformation process. The maturity model is an important tool for describing existing production capabilities. Maturity can generally be defined as the state of the complete development of things, which is the evolutionary process from the initial stage to the project stage. Currently, there is no universal measurement standard for the field of Industrial Intelligence. Moreover, the existing maturity measurement criteria are usually associated with the measurement of the degree of IT, which does not involve the degree of intelligence. However, digitalization is one of the premises of intelligent transformation, and the readiness evaluation is of great significance to the realization of Industrial Intelligence. We will introduce three methods to measure the degree to which enterprises embrace IT systems, which provide a basic judgment method for the transformation of enterprise intelligence.

DREAMY (Digital Readiness Assessment Maturity model)

The DREAMY [1] model for measuring digital maturity has two main objectives: the first is to evaluate the readiness of manufacturing enterprises to start digital transformation; the other is to identify the strengths, weaknesses, and opportunities of manufacturing enterprises and design a digital road map (Table 5.1).

If we decide to use the DREAMY method to evaluate the degree of digitalization in an enterprise, we need to identify the relevant manufacturing operation processes. Generally speaking, manufacturing business processes are divided into five main areas: design and engineering, production management, quality management, maintenance management, and logistics management. Each process area can be regarded as an independent module. For different enterprises, the areas that need to be considered can be added or deleted according to their specific circumstances. In addition,

© Shanghai Jiao Tong University Press 2020
J. Lee, *Industrial AI*,
https://doi.org/10.1007/978-981-15-2144-7_5

119

Table 5.1 The DREAMY model of maturity assessment [1]

Maturity Level 1 Initial	The process is poorly controlled or not controlled at all, process management is reactive and does not have the proper organizational and technological "tools" for building an infrastructure that will allow repeatability/usability/extensibility of the utilized solutions
Maturity Level 2 Managed	The process is partially planned and implemented. Process management is weak due to lacks in the organization and/or enabling technologies. The choices are driven by specific objectives of single projects of integration and/or by the experience of the planner, which demonstrates a partial maturity in managing the infrastructure development
Maturity Level 3 Defined	The process is defined with the planning and the implementation of good practices and management procedures. The management of the process is limited by some constraints on the organizational responsibilities and/or on the enabling technologies. Therefore, the planning and the implementation of the process highlights some gaps/lacks of integration, information exchange, and ultimately interoperability between applications
Maturity Level 4 Integrated and Interoperable	The process is built on information exchange, integration, and interoperability across applications; and it is fully planned and implemented. The integration and the interoperability are based on common and shared standards within the company, borrowed from intra- and/or cross-industry de facto standards, with respect to the best practices in industry in both perspectives of the organization and enabling technologies
Maturity Level 5 Digital-oriented	The process is digital oriented and is based on a solid technology infrastructure and on a high potential growth organization, which supports—through pervasive integration and interoperability—speed, robustness and security in information exchange, in collaboration among the company functions and in the decision making

the skills of individuals should be viewed as another dimension of analysis when assessing a company's capabilities, since deploying an intelligent manufacturing system requires expertise.

Smart Manufacturing System Readiness Level (SMSRL)

SMSRL [2] is an index that measures the degree to which a manufacturing enterprise has become intelligent. Intelligence defined by this method is essentially the intensive use of information and communication technology to improve the performance of a manufacturing system. SMSRL focuses on the maturity of the software/platform for enterprises to introduce information systems, such as Supply Chain Management (SCM), Enterprise Resource Planning (ERP), Digital Manufacturing (DM), Product Life-cycle Management (PLM) and Manufacturing Execution System (MES). SMSRL focuses on the maturity of productivity improvement, information software support, and information sharing capabilities.

5.1 Assessment of Basic Capability Maturity During … 121

Table 5.2 Maturity levels of manufacturing operations

Level 0	There is no evaluation method
Level 1	Ad hoc performance—Business operations are in the initial stage. There is no documentation or formal management
Level 2	Planning and control—Business operations are normally recorded, and manufacturing execution results are repeatable
Level 3	Standardized processes—Business operations are defined by written standards and their execution may be supported by software tools. Unexpected, abnormal events can be handled well
Level 4	Predictable performance—Business operations are defined at all organizational levels and documented. They are highly repeatable and monitored by software systems
Level 5	Continuous improvement—Business operations focus on continuous improvement and optimization

Manufacturing Operations Management (MOM)

The purpose of MOM [3] is to assess the overall situation of manufacturing facilities and to determine the organization of the manufacturing management policies, processes, and execution, as well as their reliability and repeatability. MOM does not provide a measure of production complexity, but instead a measure of process capability, especially the ability to respond to abnormal events. MOM focuses on four main process areas:

1. Production operations management
2. Inventory management
3. Operations management of quality testing
4. Maintenance operations management

Each process area consists of multiple activities including scheduling, dispatch, execution management, resource management, definition management, data collection, tracking, and performance analysis. Each activity can have a maturity level from 0 to 5. Maturity levels are represented in Table 5.2.

The higher the maturity level, the higher the organizational efficiency, which results in fewer problems that occur at the manufacturing operations management level. Maturity levels can also be applied in different areas, such as to roles and responsibilities, succession planning and backup, policies and procedures, technologies and tools, training, information integration, and key performance indicators (KPI). The disadvantages of this model are that users need to answer 832 questions before it can be completed, and there is a lack of results-based improvement strategies. However, the model can provide benchmarks for comparison with other industries and help with understanding where improvements need to be made.

The above three methods involve users answering a series of questions. From the score for each question and the total score, the information maturity level of an enterprise can be measured. In the field of Industrial Intelligence, similar tools and methods are needed to integrate the evaluation of an enterprise's embrace of IT systems and intelligence maturity, thus creating a comprehensive evaluation of an enterprise's intelligent transformation stage, so as to provide an empirical basis for enterprise decision-making.

Within Foxconn, we have also established a set of evaluation methods to measure the degree to which a manufacturing system has undergone intelligent transformation, which are evaluated from four technical elements of Industrial AI: data technology (DT), platform technology (PT), analytic technology (AT), and operations technology (OT). For Industrial AI to be successful, real-time data from the production process should be collected through sensors, programmable logic controllers (PLC), and other systems, and be combined with the mechanism of the Industrial Internet of Things (IIoT) to effectively develop Industrial AI infrastructure. However, industrial data is the basis for forming the basic information for intelligent manufacturing. Only with good quality data sets can real-time data monitoring and large data analysis technology be used to optimize the resource allocation for the production environment and predict equipment health. This can improve the transparency and performance of plant-side risk control and management, and can ultimately achieve zero failures as well as the goal of optimized intelligence and worry-free production. When managerial leaders are transforming their factories, technology is important, but only when they can evaluate their own shortcomings and optimize resource allocation strategies can they truly understand their shortcomings and self-reflect.

The maturity of DT depends on the key factors of intelligent manufacturing. The basic idea of intelligent manufacturing is to make use of industrial data collected during the production process and the technology of Industrial AI to realize intelligent design, production, and service decision-making. Therefore, data acquisition is crucial for subsequent data analysis, modeling, improving yield and efficiency, and the development of intelligent manufacturing. The maturity evaluation for data acquisition can be divided into four stages, starting from the stage where relevant data has not yet been collected, and continuing to the stage where the machine parameters can be automatically modified according to data acquisition and the established correlation model. Collecting key data that is both effective and of high quality is the core of solving problems. However, the purpose of AT is to solve and avoid problems through data, or to enable preventive measures of maintenance. Therefore, modeling, data cleaning, feature extraction, and other processing is needed beforehand, which is both critical and time consuming.

The maturity evaluation for AT can be divided into several stages, starting with the field experience and accumulated knowledge of engineers at the beginning of the project. Next are the rules derived from that problem-solving experience. Afterwards is the self-generated knowledge, which can self-evaluate the state of the model. The last stage is the ability to self-study, update, and revise the model. Based on different requirements, different tools are used for analysis. When necessary, PT and DT technologies are needed to integrate and collect multi-source data to optimize the accuracy or interpretability of the model.

OT is used to carry out the overall management and control of people, machines, materials, systems, and processes on the factory end. Its purpose is to bring about the transformation from experience-driven production to data-driven production covering all levels. The accumulation and transmission of industrial knowledge transforms

5.1 Assessment of Basic Capability Maturity During …

tacit human knowledge into a machine language which can then be converted into an executable knowledge system. This knowledge system will then use management methods to control the production and equipment operation of the factory, so as to meet the economic indicators of the factory and complete the delivery of products to the customer's satisfaction.

IT readiness is one of the preconditions of, and is of great significance to, the realization of intellectualization. The evaluation of operations technology maturity is divided into four stages, starting with operator experience rules and going on to intellectualization. On the basis of PT, DT, and AT, the intelligent model is implemented to optimize resource scheduling and processes, and to assist with the intelligent transformation of operations strategy and management methods, which provides a basic judgment method for enterprise intelligent transformation.

In the field of Industrial AI, through the integration of DT, AT, and OT maturity evaluation tools (Fig. 5.1) and comprehensive evaluation transformation stages, we can conclusively determine the level of Industrial AI development, provide an empirical basis for enterprise decision-makers, and continue to optimize, which can better assist enterprises to accurately invest resources to address their deficiencies.

Within Foxconn, the model shown in Fig. 5.2 is used to evaluate the internal advancement of Industrial AI capability readiness based on three dimensions: talent, technology, and tools. In terms of talent ability, there are five aspects: the ability to find data, the ability to use data, the ability to use tools, the ability to find insights, and the ability to change the domain from the project to the system. In terms of production system technology and management systems, we use a CPS 5C framework to evaluate the integrity of systems construction. In terms of management culture, we hope to change the previous experience-based management mode and gradually shift to an IT-based system that utilizes data-driven, real-time, remote systems integration and predictive and evidence-based intelligent management.

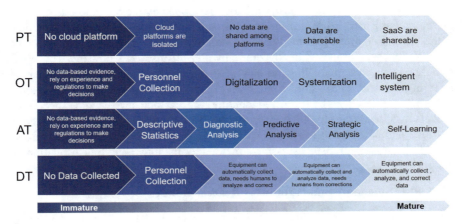

Fig. 5.1 Maturity assessment of key technologies in Industrial AI

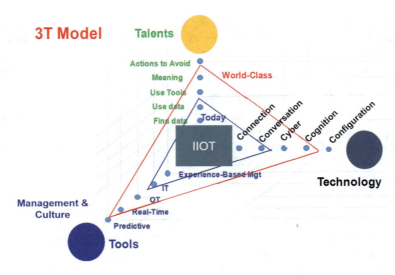

Fig. 5.2 Foxconn's Industrial AI capability readiness assessment

5.2 Assessment Tools for Global Industrial AI Enterprise Transformation Achievements

In the last section, three methods were introduced to measure the maturity of the basic capability of enterprise IT capability, which can provide the basis for assessing the transformation of an enterprise. At present in the manufacturing industry, there is no standard for evaluating intelligent capability maturity levels for industrial transformation. However, there are two well-known tools for evaluating the maturity of Industrial AI: one is the Industry 4.0 tool developed by the Singapore Economic Development Board (EDB) for the transformation of national enterprises; the other is the World Economic Forum (WEF). A maturity evaluation system for intelligent factories has been proposed in cooperation with McKinsey & Company. The WEF has organized a system to allow for enterprise evaluation in the international manufacturing industry. The system screens out exceptional models of intelligent manufacturing factories layer by layer, and identifies those which have become successful models for global Industry 4.0 transformation. This chapter briefly introduces these two international maturity assessment tools for intelligent manufacturing enterprises, providing a more pluralistic assessment method for the intelligent transformation of enterprise.

Method 1: Singapore Smart Industry Readiness Index

The Singapore Smart Industry Readiness Index (SIRI) [4] was developed by the Singapore Economic Development Board (EDB) and Technischer Überwachungsverein Süd (TÜV SÜD), and validated by an expert advisory panel.

5.2 Assessment Tools for Global Industrial AI ...

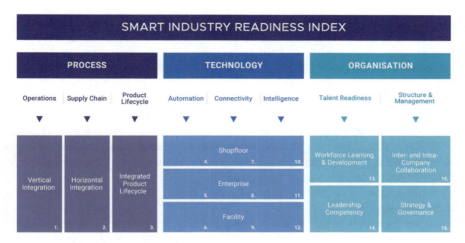

Fig. 5.3 Singapore Smart Industry Readiness Index (SIRI) [4]

It is the world's first government-developed Industry 4.0 tool for national transformation. This index will help companies determine where to start, how to expand, and how to sustain their efforts. Based on SIRI as shown in Fig. 5.3, the evaluation framework consists of three core components: process, technology, and organisation, all of which must be considered for the intelligent transformation of any facility to occur. The three cores can be divided into eight key categories and mapped according to sixteen evaluation dimensions. Companies can use these dimensions to evaluate what stage they are at on their journey towards Industry 4.0.

At the level of process change, in the context of Industry 4.0, we are no longer only focusing on process improvements that reduce costs and shorten product time to market. The concept of process has been extended to include the integration of company operations, supply chain, and product life cycle. The maturity of change at the operational level depends on whether companies adopt new technologies and methods to achieve faster conversion of raw materials and labor into goods and services at the lowest cost. The change in the supply chain means that the traditional supply chain model is becoming more and more digitized. The key data are collected through sensors loaded within the supply chain and connected to the central data center for analysis and management. Data can also be gathered during the product's lifecycle. Business systems and personnel can be better integrated, creating a unified information platform that can be digitally managed, allowing decision makers to more effectively monitor the whole process across the entire value chain.

At the level of technological change, through technologies such as cloud computing, machine learning, and the Internet of Things (IoT), an intelligent industrial environment that integrates physical assets, equipment, and enterprise systems can be constructed in order to achieve continuous feedback and analysis of dynamic data. This integrated system can help create a company more flexible and agile. Its scientific and technological change evaluation falls into three categories: automation, connectivity, and intelligence. Automation is the application of technology to monitor, control, and implement production and delivery related to products and services. The process is more flexible than before, so production planning can be determined according to the demand of small-batch production in a cost- and time-effective way. Connectivity is the measure of the state of interconnection among devices, machines, and computing systems so that they can communicate with each other. Intelligence is the ability to process large amounts of real-time data through cloud and data analysis technology, and summarize them into decision-making suggestions for subsequent operations. This allows for problem diagnosis and a better understanding of how to improve the system. Through machine learning, intelligent systems can predict both equipment failures and changes in demand patterns in advance. In the face of changing internal and external business needs, intelligent systems can even make decisions independently.

At the organizational level, the company must adjust its organizational structure and communication process in order to convey the transformative mission of Industry 4.0 to the entire staff and keep up with the pace of change. The first key to achieving this goal is in the composition of the organization. The second key is the way in which the organization operates. As organizations adopt a flatter structure, it becomes critical for companies to build a flexible team to allow for continuous learning and development. Organizations need to focus more on decentralized management, move the power center downward, and improve decision-making democratization and efficiency. Through information sharing, more cooperation opportunities will be created with internal and external partners.

Companies can use sixteen dimensions to evaluate their status during the transformative process of Industry 4.0. As shown in Table 5.3, each dimension examines the current process, system, and structure.

SIRI provides a systematic method for evaluating the company's Industry 4.0 transformation. As an enterprise management system, we should attach importance to the direction, timing, and degree of transformation of intelligent manufacturing. Although the relative importance of each company's three core components, eight key categories, and 16 dimensions will not be the same, SIRI provides a common language that companies can use to promote coordination within the company and cooperation and innovation outside the company.

Method 2: The WEF's Intelligent Factory Maturity Assessment

The World Economic Forum (WEF) is a non-profit organization which was founded in 1971 and is headquartered in Coloni, Geneva, Switzerland. Every year, the WEF gathers global leaders from business, politics, academia, the media, and other fields to discuss the most pressing issues facing the world. The theme of the 2017 Forum was the Fourth Industrial Revolution in manufacturing. The organization and McKinsey & Company collaborated to develop a maturity evaluation system for intelligent factories, which was used to select global model factories in line with the Fourth Industrial Revolution.

5.2 Assessment Tools for Global Industrial AI …

Table 5.3 Sixteen dimensions of SIRI [4]

Dimension 1	Process—Vertical Integration	Establishes end-to-end data threads by digitally managing processes and systems at all levels in a factory
Dimension 2	Process—Horizontal Integration	Refers to the integration of processes within an organization and stakeholders outside the organization
Dimension 3	Process—Product Life Cycle	Integrates people, processes, and systems throughout the product life cycle
Dimension 4–6	Automation—Shop Floor, Enterprise, and Facility	Assesses the degree of automation and flexibility, as well as integration, across multiple systems at workshop, enterprise, and plant levels
Dimension 7–9	Connectivity—Shop Floor, Enterprise, and Facilities	Assessment of connectivity between equipment, machines, and systems installed at the workshop, enterprise, and facility levels
Dimension 10–12	Intelligence—Shop Floor, Enterprises, and Facilities	Systems at workshop, enterprise, and plant levels can identify and diagnose any deviations and adapt to changing requirements
Dimension 13	Organization—Workforce Learning and Development	Aims to build organizational excellence by cultivating the ability and skills of the labor force
Dimension 14	Organization—Leadership Competency	Adopts the latest concepts and technologies to maintain the competitiveness of the company
Dimension 15	Organization—Inter- and Intra-Company Collaboration	The process of collaboration between internal and external partners to achieve a common vision and goals
Dimension 16	Organization—Strategy and Governance	Designs action plans and implements them to achieve a series of long-term goals

Based on the WEF's factory maturity assessment, as shown in Fig. 5.4, the maturity of factory change can be assessed using four project categories: technology application, change cases, impact benefits, and contributing factors. The maturity of each project can be divided into three levels: the initial stage, the developing stage, and the advanced stage. The more projects that meet the criteria, the more the factory aligns with the WEF's Model for Factory Standards for the Fourth Industrial Revolution. Such companies are pioneers of global change as well as a models for excellence in enterprise.

	Dimension	Maturity Level		
		Initial Phase	Developing Stage	Advanced Stage
Effect	Financial Effect	<10%	10-30%	>30%
	Operational Effect	Small Changes		Obvious Changes
Case	Transformation Case Amount	1	2-3	≥4
	Case Scale	Only demo area use 4IR transformation case	Specific areas use 4IR transformation case	4IR transformation case used in all factory relative areas
Technology	4IR Tech Platform	No focus on the scalability of tech platform		Scalable tech platform and 4IR technologies
	4IR Tech Amount	1	2	≥3
Promoting Factor	Organization Structure & Management	Allocation of resources and authority is not clear	Allocation is based on project	Expert teams and clear authority
	Talent & Ability	No Industry 4.0 Talent	Has Industry 4.0 Talent	Non-traditional talent management system, and training courses
	Engineering Development	Waterfall Development Model	Agile Development Model	Frequently Overlapping Agile Development Model
	Community System	No Industry 4.0 Community	Company has internal Industry 4.0 Community	Company has interdepartmental Industry 4.0 Community, like university and industry cooperation center
	Management & Communication	High level leaders have little participation	Broadcast transformation stores and set up performance management	Has transformation stories, highest leaders serve as a model, establish clear performance management and reward mechanism
	Document Filing	Central productions system has no document filing	Central productions system has document filing	Complete production record, easy to retrieve files and are updated periodically

Fig. 5.4 World economic forum intelligent factory maturity assessment [5]

First, at the level of technological evaluation, the standard considers whether the factory has adopted an intelligent technology platform to assist production and monitoring. Intelligent systems are used to track production line status in real time, collect huge amounts of data, analyze data, or provide decision-making suggestions. If the plant is not developed or the platform is not used to monitor production line status, the project will receive a lower maturity score. If the factory uses an intelligent platform for monitoring, forecasting, warning, analysis, decision-making suggestions, and so on, the project will get a higher maturity score.

Secondly, at the case evaluation level, the extent to which the plant has adopted reforms is examined, and the reforms are scored according to the number of projects and the attributes of the plant area. A higher maturity score is awarded when the factory carries out a variety of reform projects, depending on the execution field of the project. If the reform project is tested only in the demonstration production line, a lower maturity score is awarded. If the reform project is applied in the actual production line, a higher maturity score is awarded. The following are some plant change projects for reference including, but not limited to: predictive maintenance through audio, temperature montioring, machine vibration monitoring, cost optimization of complex operation through sensor analysis, real-time components positioning system, digital work standardization through mixed reality, and systems integration. The system platform is integrated to achieve remote monitoring and production optimization, and the machine alarm system is used to make automated logistics operation decisions.

Third, at the level of impact assessment, we examine whether the operational and financial status of the production line or factory can be greatly improved after the implementation of transformational technology—for example, reduction in costs and inventory, improvements in quality, efficiency, and so on.

Finally, contributing factors will be used to evaluate the organization's multiple projects, including organizational structure, organizational management, talent cultivation, engineering development, communities, organizational communication, document filing, and so on. In the early stages of reform, the factory may not have a clear organizational change framework or may be lacking Industry 4.0 related personnel or training systems. It may also be the case that the change action plan is only discussed by the management of the organization or that no documents have been filed. On the contrary, factories with a high degree of transformation have clear organizational change frameworks, complete Industry 4.0 talent training systems, effective communication from top to bottom, and clear positioning and goals for everyone involved. They adopt continuous iterative agile development systems, complete production records and systematic archiving, etc. Factories with some of these characteristics will receive higher maturity ratings.

The two methods mentioned above are used to evaluate and score enterprises or organizations comprehensively, including their operations, supply chain, workshop, talent base, science and technology, management methods, and so on. Enterprises or organizations can evaluate which stage of industrial transformation they belong to, assess the maturity of transformation and the direction for improvement, and then provide decision makers with the basis for judging the transformation of intelligent manufacturing. Enterprises must immediately launch smart manufacturing transformation actions, regardless of the size or nature of the company, in order to benefit from Industry 4.0 transformation.

5.3 Foxconn Lighthouse Factory

At the World Economic Forum (WEF) in January 2019, McKinsey, in collaboration with the forum's sponsors, selected 16 "lighthouse factories" from around the world to honor the benchmark leaders who had successfully adopted and integrated the cutting-edge technologies of the Fourth Industrial Revolution. A white paper entitled "Fourth Industrial Revolution: Beacons of Technology and Innovation in Manufacturing" [6] was published at the same time, which provides an in-depth interpretation of these lighthouse factories and discusses their key characteristics and success factors (Fig. 5.5). From the first-hand perspective of factories, senior managers, technology pioneers, and industry chain stakeholders, key value drivers and expansion drivers are highlighted, which make these pioneers of the implementation of intelligent manufacturing different.

Foxconn's lights-out factory has also been honored as a lighthouse factory (Fig. 5.6). It should be noted that "lights-out factory" does not refer to a specific factory, but a new standard and manufacturing system defined in the factories that produce various products. It also includes the manufacturing of die and mobile phone structural parts, as well as SMT production lines for circuit boards and semiconductor factories for the most advanced 8K display panels. These factories have the same

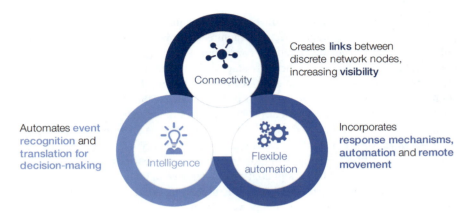

Fig. 5.5 The Three characteristics of lighthouse factory [6]

characteristics—that is, they have interconnectivity, intelligence, and reconfigurable automation capabilities. Below are the three characteristics of a lighthouse factory.

- Smart connectivity: a network of data and edge computing between different manufacturing devices and nodes to enhance the transparency and collaboration of the system.
- Flexible automation: combines reconfigurable response mechanism, automated execution, and collaboration, and supports remote configuration and task delivery. It enables manufacturing systems to quickly respond to customization requirements and has strong resilience to internal and external interference.
- Intelligent Prediction and Decision-Making System: a closed-loop predictive analysis system that combines large data and intelligent algorithms for automatic event recognition, impact assessment, and decision optimization to predict invisible problems.

In Foxconn's lights-out plant, we focus our energy and technology on three "W" jobs: waste reduction, work reduction, and worry reduction. These tasks sound simple, but they have gone through a long process of accumulation and exploration. By means of process improvement and continuous perfection of digitalization early on, we also continuously optimize efficiency and quality control to eliminate unnecessary downtime and quality defects, constantly excavate the comprehensive efficiency and cost optimization of manufacturing systems, and solidify the standards and experiences to form standard rules in the digital system. Furthermore, we use new paradigms, processes, and real-time monitoring of current production efficiency to allow for optimization. In the benchmarking of cluster optimal practices compared with historical optimal practices of systems and different manufacturing units, the space for dynamic evaluation and efficiency optimization is sought. This is the process of waste reduction. Through the integrated automation production line, almost everyone's work has been replaced in some factories, and with the material transportation in the workshop, the flow of products between different production units

5.3 Foxconn Lighthouse Factory

Fig. 5.6 Foxconn was selected as a world economic forum lighthouse factory

and the quality inspection have been automated. More importantly, Foxconn's lights-out plant has invested a lot of design considerations in the balance between integrated automation and flexibility. Especially for the fast iteration of consumer electronic products, the efficiency climb and yield stabilization speed in the process of New Product Introduction (NPI) are in the absolute leading level of the industry. This requires not only deep experience, but also the quick execution of software and hardware design. Work reductions are not just replacing people with automation, but also making the production line as flexible as people, so that people's experience and knowledge can be applied and integrated into the system.

If waste and work reductions are areas that Foxconn is working on and has made achievements with in the past, then what we now need to further create is worry reduction derived from carefree operation. The significance of lights-out factories is not just that no one in the factory needs to turn on the lights, but that when we turn off the lights, we do not need to turn them back on.

The key to realizing a worry-free manufacturing system is to make the previously invisible problems explicit, to manage the root causes of the problems and the process side, and to change the way of solving visible problems by correcting errors and avoiding problems through accurate predictions. For example, if we are worried about the yield of products, we analyze the relationship between process parameters to understand the reasons for their impact on yield, and then establish an early warning model for abnormalities in the process parameters to avoid future yield problem.

Similarly, if we are worried about the loss of production capacity caused by equipment downtime, we can achieve predictive maintenance for failure by continuously monitoring the condition parameters of the equipment, establishing a health assessment model, predicting the risk of equipment failure, and the remaining usage time. This will not only avoid the losses from downtime, but also will reduce the costs incurred by over-maintenance.

In the field of machining, we can predict the optimal replacement time of each tool by predictive modeling of the remaining useful life of machine tools. This advancement can not only improve yield from 99.4 to 99.7%, but also reduces tool cost by 16% and unexpected downtime by 60%. In the SMT process of circuit board manufacturing, we established a recession assessment and predictive maintenance model for industrial nozzles, and were able to perform health cycle prediction and automatic replacement. Predicting suction nozzle life can effectively reduce its maintenance requirements, replacement time, and cost by up to 66%, while reducing suction nozzle inventory by 64%.

I frequently reference a formula for calculating manufacturing competitiveness: *manufacturing competitiveness = quality/cost * customer value*. We need to continuously achieve higher quality with the smallest cost and create more value for customers. In Foxconn's lighthouse factory, we have both improved quality and reduced cost through predictive analysis technology. Flexible automation enables customers to innovate and iterate faster, expanding the boundaries set by manufacturing system capabilities for innovation.

Over the past 20 years, there have not been many breakthroughs in the technology and capabilities of manufacturing systems. Demand and R&D investment have become more and more dispersed. In many point applications, some enterprises have successfully crossed the pilot stage and started to promote the innovation brought about by the Fourth Industrial Revolution on a large scale in practice. They can achieve unprecedented increases in efficiency with the fewest number of employees. If enterprises and governments can work together to promote the technology of the Fourth Industrial Revolution on a large scale, the global economy may usher in new development opportunities and bring greater well-being to more people.

Foxconn's lighthouse factory is a model of both digital manufacturing and global Industry 4.0, which has all the necessary features of the Fourth Industrial Revolution. It also validates the hypothesis that the improvement of the driving factors of production value can lead to new economic value. These driving factors include resource productivity and efficiency, flexibility and responsiveness, product launch speed and customization ability to meet customer needs, improving the production system of traditional enterprises, innovating the design value chain, and creating new business models with subversive potential. All these factors can create value.

The successful application and promotion of the "lighthouse factory" proves that there is more than one way to embrace the intelligent industrial revolution as manufacturing pioneers can choose from multiple development paths when planning for the future. These paths are not mutually exclusive, but complement each other and go hand-in-hand.

- Intelligent operation of production systems: by applying automation, IoT, predictive analysis, and AI technologies to continuously improve the operational capacity of production systems, enterprises can expand their competitive advantages. They aim to optimize production systems and improve operational efficiency and quality indicators. Usually, enterprises will pilot in one or more factories, and then gradually apply protocols to other factories.

- End-to-end value chain innovation: through technological and model innovation, enterprises can create new businesses. They deploy innovation throughout the value chain, providing customers with new or improved value propositions by introducing new products, new services, high customization, smaller batches, or shorter production cycles. Firms will first implement innovation and transformation in a value chain, and then gradually extend their experience and capabilities to other sectors.
- Improving employee skills and shaping future work patterns: by improving employee understanding of new technologies such as data analysis, intelligent robots, and add-on manufacturing, the distance between these technologies is shortened and the cycle of the introduction of new technology is reduced. In this way, employees can acquire professional skills that enable them to assume new roles that have been established as a result of transformation such as "Network Security Supervisor", "Data Engineer", "Digital Lean Operations Supervisor", and other positions.

5.4 How to Construct the Organizational Intelligent Transformation Ability in Industrial Enterprises

The real digital transformation menu is limited to the application and implementation of new technology, which requires the coordinated transformation of enterprise strategy, talent, organizational models, and business models. The biggest challenge faced by organizations in the pursuit of digital transformation is to find, train, and retain suitably skilled workers.

According to the global AI talent report from LinkedIn, a leading global talent community, the competition between enterprises driven by transformation data technology will eventually be upgraded to competition for core talents. Enterprises are paying more and more attention to the introduction of core talent, but also face enormous cost pressure. The salary level of core talent in the field of advanced technology such as big data/AI is much higher than that of traditional skilled IT workers, especially in the field of industrial big data. The lack of knowledge pertaining to the industry as a whole and specifically towards AI has led to the dilemma of companies not being able to find, recruit, or retain large numbers of big data scientists. Among the six driving factors of manufacturing competitiveness proposed by the Deloitte Global Manufacturing Competitiveness Index, talent is recognized as one of the most important.

American Industrial Research Talent Training-School-Enterprise Cooperative

In terms of personnel training, besides investing significant resources into basic scientific research, the United States also pays special attention to the integration of industry, universities, and research, as well as the training of industrial research talent. There is a long-standing problem within American enterprises: when inventions and creations reach a certain level of basic research, they often suffer setbacks when they are industrialized on a large scale. It is not just technology or investment that prevents these inventions from reaching industry. Industry is often not ready to accept these new innovations, and they are often beyond the scope of academic laboratories. This phenomenon is called the "Death Valley Curve" of innovation (Fig. 5.7).

In order to help the technological innovation of academia and business cross this "Death Valley Curve," the National Science Foundation (NSF) launched the Industry/University Cooperative Research Centers (I/UCRC) Program in 1973, aimed at developing long-term partnerships between industry, academia, and government (Fig. 5.9). The NSF invests in these partnerships in order to promote research projects with common interests, helping to improve the innovative capacity of engineering and scientific workers, and promoting the transfer of technology to industry through the integration of research and education on the basis of national research infrastructure. The NSF also encourages international industry-university cooperation and promotes these goals globally: the mode of cooperation between industry-university cooperative research centers and enterprises is a membership system, i.e. member enterprises invest a fixed amount of money each year into research areas of common interest to all member enterprises, and the research results are shared with all the member enterprises. Sharing between members can be likened to crowdsourcing for scientific research, where an Industrial Advisory Board (IAB) is convened every six months.

Participants in IAB meetings report on research results and discuss new research interests. In principle, the I/UCRC Program's research center investments are broken into 5-year stages. The research center qualification in a field can be used for

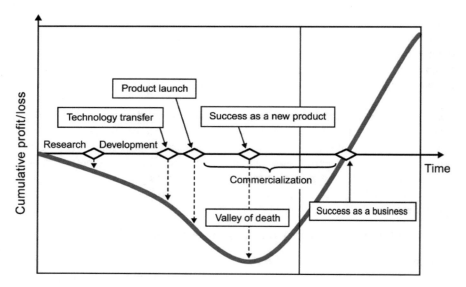

Fig. 5.7 "Death Valley Curve" of innovation [7]

up to 10 years. After the expiration, the research results should be invested in within the industry by spin-off companies or technology authorization. This means that the research center has graduated from the project and the technology has been successfully commercialized. Today, the I/UCRC Program has successfully funded over 100 research centers, touching upon such fields as 3D laser metal printing, artificial biological organizations, computer-aided manufacturing, wireless sensors, and communication networks. These efforts have been funded by the I/UCRC Program in today's influential technical fields. Most of the graduates from these research centers will enter into enterprises and become excellent industrial research talents. They have strong scientific research and innovation ability, and have received systematic product and market strategy training. Their performance in enterprises is very prominent.

I initiated the IMS Center under the I/UCRC Program in 2000, and since that time it has developed into an alliance of four universities including the University of Cincinnati, the University of Michigan at Ann Arbor, the University of Texas at Austin, and the Missouri University of Science and Technology. Since its inception, the center has worked with more than 100 companies throughout the world, and nearly 300 funded research projects have been completed. Eighty percent of the graduates work for member companies, including industry leaders such as GE, Bosch, Tesla, and Boeing. In 2012, the NSF assessed the economic impact of all I/UCRC research centers. The IMS Center ranked first among all research centers with an economic value of $850 million and a 1:238 return on investment ratio. In addition, with the technology developed by the IMS Center, its researchers have successively achieved excellent results in the annual industrial data analysis competition held by the PHM Society, and the IMS is known as the "West Point Academy" in the field of industrial big data. From the perspective of IMS training systems, we integrate industrial big data into the engineering industry, and divide the curriculum into four main areas: data technology (DT), analytic technology (AT), platform technology (PT) and operations technology (OT). The vision of the course is the achieve "the ability to find data", "the ability to use data", "the ability to use tools", "the ability to analyze data", and "the ability to develop insights". The IMS Center believes that only when cross-domain technologies are integrated can industries learn from each other, and only when industries learn from each other, can more general technologies be produced.

According to my many years of experience in personnel training, the training of industrial research talents can be divided into four "P" stages (Fig. 5.8), namely Principle (theoretical research), Practice (scientific research practice), Problem-solving, and Professional. In its theoretical research training, IMS pays attention to the cultivation of thinking frameworks, algorithm theory, industrial knowledge and modeling tools, and mainly develops this knowledge through course training. In practice, the IMS Center often deals with real problems and real data from industry. Such data sets are often designed to create training competitions similar to the big data competitions hosted in industry, which encourage the use of learned knowledge to solve a clearly defined problem, and require students to improve assigned technical indicators. Competing with each other, the results of these advances are then incorporated

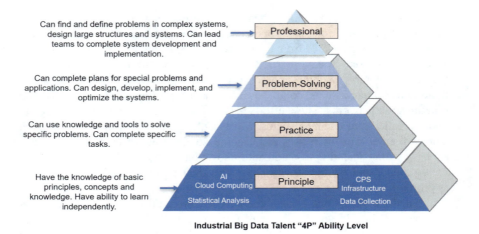

Fig. 5.8 Pyramid of industrial research talents training

into student research papers, which often focus on how to solve a specific technical bottleneck and form a useful reference for the industry. In Problem-Solving, students at the IMS Center spend at least three months each year interning in member enterprises to complete an entire project design, function development, system implementation, and optimization process. These internships tend to become student research funding projects, often lasting about two years, and may be expanded upon for student graduation thesis projects. After completing the above training, students also need to cultivate the ability to discover and define problems in complex systems, to design large-scale schemes and systems, and to lead teams to complete system development and implementation. These abilities need to be continuously cultivated in their work, and encourage them to gradually grow into industry-oriented research leaders.

Regarding the basic maturation of the industrial intellectualization transformation mentioned in the previous section, we evaluated it based on four aspects: DT, AT, PT and OT. Correspondingly, for an enterprise that wants to seek industrial intelligent transformation, DT, AT, PT, and OT talents are indispensable. The company should also have a standard evaluation index for talent classification and evaluation. Big industrial data has three main characteristics: fragmentation, low quality, and background, which also make up the basis for industrial data. Therefore, DT data processing technology is very important to solve these problems, but the application of DT technology requires some background information, including needing to understand the engine, electronic manufacturing, and other areas of knowledge; AT needs to use computer science and artificial intelligence and other computing technologies; PT can produce knowledge, not only to share but also to give feedback to OT.

At present, many colleges and universities have established DT, AT, PT, and OT related courses and conventrations, but there are still many bottlenecks in the course construction. First, industrial data science coursework requires the incorporation of data analysis, especially management platforms for industrial big data. Second, there is a lack of applied data. Because of the particularities in data science, enterprises have a hard time in colleges and universities. There is more enterprise data and less university data. Third is the lack of practical experience stemming from schools and enterprises rarely cooperating. Fourth, it is not the superposition of these courses or the integration of talents, so the current industrial AI training lacks an integrated platform running through the cooperation of data education institutions and enterprises. The ultimate goal of industrial big data is to create value. Only under the premise of creating value and on the basis on the industrial scene of big data can these four aspects of technology and knowledge be integrated into the overall training system. Only this sort of integration can be considered a systematic integration.

The training of Industrial AI requires theoretical study, practical platforms, and the integration of industrial development. Industrial AI is not like "finding nails with a hammer", but requires understanding the real pain points in production from the field. The integration of theory and practice is the key to personnel training. Universities should go out to find more fields and data. Enterprises should open their doors and let data scientists from all over the world solve problems in production.

5.5 Open Source Industrial Big Data Competitions

In recent years, industry-related big data modeling and analysis competitions have attracted more and more attention from academics and business researchers. Since the first industrial big data competition was held by the Prognostics and Health Management Society (PHM Society) in 2008, there have been more data modeling competitions for the industrial field organized by the PHM Society, IEEE-PHM Society, Kaggle, etc. These data competitions not only provide very real problems and data in industrial scenarios, but also provide a good platform for different methods to communicate and compare with each other. Taking the 2008 industrial big data competition as an example, the aero-engine life cycle data and remaining useful life prediction provided by General Electric (GE) has become the classic case study for hundreds of papers and nearly 100 doctoral dissertations.

This chapter contains nine industrial big data challenges in the United States from 2008 to 2017 and summarizes the difficulties of the challenges and the thinking of the winners (Table 5.4). The competitions cover a wide horizontal range of industries including aerospace, rail transportation, wind power, machine manufacturing, semiconductor manufacturing, and other industries, while spanning different vertical monitoring levels.

In addition, it is worth mentioning that unlike Internet big data modeling, it is difficult to find open source modeling data in industrial scenarios. All the data sets and detailed problem-solving papers collected in this chapter can be downloaded on the Internet, which will be helpful for readers to learn and practice in the field of industrial intelligent modeling.

Table 5.4 List of industrial big data challenges

Completion	Industry	Monitoring level	Data characteristics
Remaining useful life estimation of aircraft engine	Aviation	Components	Training samples are few, monitoring variables are many. Data includes life cycle data of components. Dimension reduction method is needed
Fault detection and diagnosis for gearbox	Rotary machinery	Components	No Label. Modeling is dependent on vibration analysis
Remaining useful life prediction for CNC milling machine cutters	Machining and manufacturing	Components	Because of the strong dependence on working conditions and the great influence of noise, it is necessary to combine the physical model with the data-driven method for modeling
Anemometer health assessment	Wind power	Components	Greatly affected by working conditions and requirements
Remaining useful life estimation of ball bearings	Rotating machinery	Components	Modeling is dependent on vibration analysis
Asset risk assessment in industrial remote monitoring	Electronic manufacturing	Equipment	The meaning of data is missing, and noise has great influence on modeling. Data-driven modeling is needed
Fault detection and prognostics in industrial plant monitoring	Power production	Fleet	The working conditions and environments have great impacts on the model, and data meaning is absent. It is necessary to adopt data-driven modeling method
Virtual metrology of chemical mechanical planarization for semiconductor manufacturing	Semiconductor manufacturing	Equipment	Working conditions and environments have great impacts on the model, and meaning of data is absent

(continued)

5.5 Open Source Industrial Big Data Competitions

Table 5.4 (continued)

Completion	Industry	Monitoring level	Data characteristics
Fault detection and diagnosis of train bogies	Rail transportation	Components	Working conditions and environments have great impacts on modeling. Mixed modeling can be used

2008 PHM Data Challenge: Remaining Useful Life Estimation of Aircraft Engine

The problem in the 2008 PHM competition was to predict the remaining useful life of an aircraft engine based on historical data, which came from the commercial modular aero-propulsion system simulation aircraft (C-MAPSS) [8] developed by the National Aeronautics and Space Administration (NASA) [9]. C-MAPSS is a high-precision engine calculation model. Through adjustable parameters such as environmental variables and operating conditions, the controller parameters can simulate the operating parameters of an aircraft engine under different operating and health conditions. In this data competition, participants were asked to analyze the simulation data generated in C-MAPSS to estimate the remaining useful life of the engine (Fig. 5.9)

The data in this competition was divided into a training set and test set, with 218 samples each. The training data gave the life cycle data of the aircraft engine, while the test data only provided engine degradation data. The requirement for the competition was to predict the remaining useful life of the aircraft engine based on the training data. The original data of the competition simulated many process parameters of the aircraft engine under different operating conditions. The initial deterioration and system errors of each engine were unknown, and the data length of each sample was also different. Challenges in this competition mainly included the following aspects: the operating conditions of the aircraft engine were complex, and it was difficult to predict the remaining useful life of an aircraft engine; secondly, the data was a multivariate time series, which contained the differences in working conditions and noise. When dealing with the data, it was difficult to manage the data with a higher dimension which needed to be considered. Useful signals in the original data were screened by correlation and other indicators.

The championship team in this competition was made up of Tianyi Wang and others from the IMS Center in Cincinnati [10], who used a method to predict the remaining useful life based on the similarity of engine degradation trends. The first step of this method was to classify the data according to the working conditions, and preliminarily screen the useful data by using the monotonicity of equipment degradation. The second step was to establish a logistic regression model based on training data, thus mapping the process parameters to the engine degradation trend. Based on the trained health model, the degradation trend of all aircraft engines in the training data was calculated and a training sample database was established. Finally, by comparing the similarity of the engine degradation trend between the test data

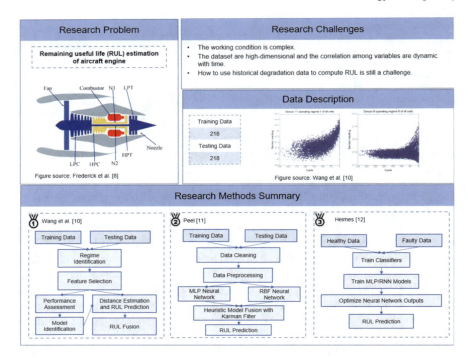

Fig. 5.9 2008 PHM data competition—remaining useful life estimation of aircraft engine

and training data, the engine health status was reasonably evaluated in the test data, and the remaining useful life was predicted. The advantage of this model was that it was simple to use, and it gave the best prediction accuracy for the 2008 data.

The second-place team was Peel [11] from the BAE Systems Advanced Technology Centre. The method adopted was the use of a neural network combined with a Kalman filter. Similar to the previous process, the method first classified the data, classified six working conditions, and then preprocessed and normalized the data in order to present all the data at the same scale. In model training, the method randomly selected part of the training set data to build the model, and the rest of the training set was used to verify and adjust the established training model. The neural network used multi-layer perceptron and a radial basis function. However, these two models had shortcomings, so in view of the deviation between the perceptron model and radial basis function in predicting the remaining useful life, the Kalman filter was introduced to improve the method, and then the model was randomly classified by using an eliminating heuristic method. Finally, the models under each classification were compared. With the selection of each round, the proposed method was used to improve the prediction of the remaining useful life in order to find a better performance model. The team ultimately selected the best model to arrive at their calculation of remaining useful life.

The third-place team in the competition was Heimes [12] of BAE Systems. Their method also used the recurrent neural network. First, some data from the begining and end of each cycle were taken as health samples and faulty samples respectively, and a classifier was built by using a multi-layer perceptron. However, the recurrent neural network and extended Kalman filter were used to predict the remaining useful life of the engine. Recurrent neural networks use internal neurons and feedback to learn complex non-linear mapping relations, while the Kalman filter can remove data noise and follow the dynamic changes of data. In order to optimize the model, a genetic algorithm was used to improve the model. After being compared with each other, three models were selected to provide a comprehensive average, and the final prediction results were obtained.

Overall, these three methods were similar in data processing. First, the teams classified the data according to working conditions and extracted features; then, using features to establish multiple algorithm models, the teams substituted the test data and compared the performance of each model. Finally, the teams optimized the results. The results of the final competition show that the algorithm that got first place in the competition considered the similarity between the test data and the training data based on usage, and the similarity in calculation performed well in processing clustering. The other two algorithms used a neural network or operated within the framework of a neural network system, added a Kalman filter, or extended a Kalman filter to remove noise in order to improve the final prediction results.

2009 PHM Data Challenge: Fault Detection and Diagnosis for Gearbox

Another event held at the 2009 PHM Data Challenge involved the fault diagnosis and classification for a gearbox. The competition required teams to use vibration data and gear box specifications to correctly identify the faulty parts, their locations, and the degree of damage in a gearbox system [13]. The competition gearbox data set contained 560 data samples. The data acquisition of each sample took four seconds, and consisted of three data channels. Two vibration data sets were collected by two accelerometers (sampling frequency of 66.67 kHz) fixed on the support plate of the input and output axes, respectively, and the rotation speed (10 pulses per revolution) signals were collected by a tachometer installed on the input axis [14] (Fig. 5.10).

Competitors needed to make full use of signal processing methods and gearbox related expertise to judge the type and degree of gear failure without any fault labels in the data set. The original data considered two kinds of loads and five rotational speeds, and simulated 14 kinds of different gearbox faults on different fault levels.

The top two finishers in the competition were from the IMS Center at the University of Cincinnati. The first-place team in the competition used the information recombination method. First, they combined 18 bandpass filters to obtain the reconstructed vibration spectrum. Then, samples were clustered to separate working conditions and 14 fault modes were classified by holo-coefficients radar chart [15]. This method skillfully superimposed simple band-pass filters, decomposed the frequency domain for global analysis, and at the same time, had insights into the characteristics of the local frequency domain, all resulting in a good filtering effect.

142   5 How to Establish Industrial AI Technology and Capability

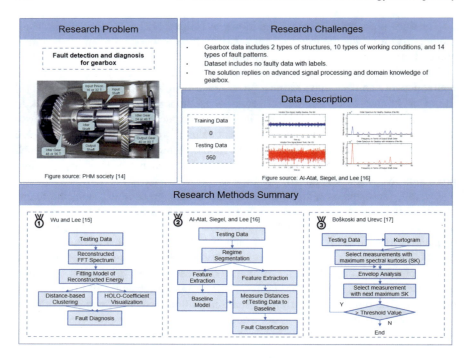

Fig. 5.10 PHM data challenge 2009—fault detection and diagnosis for gearbox

The second method [16] first extracted a large number of features (more than 200) by using common signal processing methods in the time and frequency domain, wavelet decomposition, and envelope spectrum, then separated loads by observing the change in speed of the input shaft, identified different tooth types by using gear meshing frequency and spectrum similarity analysis, and finally performed a health assessment. The fault-free samples were taken as the benchmark, and the distance between other samples and the benchmark was measured. The classification results were given through the probability model. The advantage of this method was that health benchmarks were given by the health assessment for comparison with other samples, which successfully solved the problem of the training model not being able to be trained due to the fault-free label in the competition.

The method used by the third-place finishers provided an idea for separating fault modes based on envelope spectrum and spectral kurtosis [17]. Before the envelope spectrum analysis of vibration signal, this method used a bandpass filter at a specific carrier frequency to filter the signal. The advantage of this method is that it can effectively reduce the influence of noise on useful signal components and restore the components containing fault information to a greater extent.

Generally speaking, the key to gearbox fault diagnosis lies in how to effectively use the knowledge from signal processing and physical fields to find out the special features corresponding to specific faults. The fault diagnosis for the gearbox in this competition mainly considered the fault behavior of the gearbox under stable working conditions, and the data acquisition was completed under the conditions of fixed load and fixed speed. In practical industrial applications, gearboxes often operate in more complex, dynamic conditions, and the situation of variable speed and load occurs from time to time. Therefore, in practical applications, the aforementioned methods should be improved to achieve the intelligent health management of gearboxes under complex conditions.

2010 PHM Data Challenge: Remaining Useful Life Prediction for CNC Milling Machine Cutters

The term milling machine mainly refers to machine tools that process planes, grooves, gears, and various curved surfaces through the rotary motion of a milling cutter, and are widely used in the manufacturing industry. A milling cutter is a rotating tool with multiple cutter teeth; anytime the cutter teeth accidentally fracture, it will disrupt or even delay the production plan and cause economic losses. Therefore, the importance of health monitoring and remaining useful life prediction for the milling cutter is obvious. The research objective for the 2010 PHM Data Competition was remaining useful life prediction for CNC milling machine cutters.

The milling machine for the data competition was a high speed CNC machine (Röders Tech RFM760) with a spindle speed for the milling cutter of 10,400 revolutions per minute (RPM). The milling cutter in the CNC milling machine repeated milling until the cutter teeth broke. The organizer of the competition provided data from six milling measurement categories for a single type of milling cutter across the duration of the whole life cycle. The measurement data for the three milling cutters were used as training data, and the wear measurement values for each set of cutter teeth were provided as well. The remaining three milling cutter measurements were used as test data. The goal of the competition was to predict the wear rate for the milling cutter after each milling process based on the measured data, and to predict the maximum number of milling times that the milling cutter could achieve before reaching a certain set wear rate.

There were several challenges in this data competition. First, although six milling cutters were of the same specifications, the initial wear states and wear trends for each milling cutter were different due to possible deviations in the production and installation processes and the different material densities of the milling work pieces. Second, the actual milling environment was harsh, and the data acquisition has significant background noise, which made it difficult to preprocess signals and extract features. Third, milling is a complex cutting process: many different factors affect the actual wear of the milling cutters. The competition only provided three milling cutter measurements. As training data, it was a challenge to model milling tool wear adequately.

Because of the complexity of the milling process, it was difficult to establish the physical model for the milling cutter wear process. The champion and runner-up for this competition used data modeling to predict the wear for the milling cutter. The champion team included Sreerupa Das and others [18] from the American global aerospace, defense, security, and advanced technologies company, Lockheed Martin.

Although the data analysis process was simple, effective methods were proposed for each step according to the characteristics of the milling process. In the process of data de-noising, considering the abnormal vibration caused by milling cutter contact and leaving the work piece surface in the milling process, the first and last measurements were deleted in each milling process, and the measurements for the milling process were deleted in consideration of the edge effect of the cutter teeth. When the cutter teeth wore out, the milling force needed to complete the same milling task increased, so in the feature extraction process the time domain statistical index of the X, Y, Z three-axis milling force was calculated as the feature. Considering that the milling cutter contained three cutter teeth, X, Y, and Z triaxial vibration data, the energy values on the cutter teeth passing frequency (3 times rpm = 520 Hz) and frequency doubling showed different degrees of wear for the cutter teeth.

In the model training process, a three-layer neural network was used to learn the wear trend of the cutter teeth from the extracted features. In order to optimize the training efficiency and accuracy of the traditional neural network, the weight of the neural network was adaptively adjusted by using the method of resilient back propagation. Considering that the wear on the cutter teeth was a cumulative process, the data from the last milling process was used as input for the neural network model. At the same time, considering the influence of tool life on tool wear trend, milling times were also provided to the neural network as input. Finally, Das selected one of the 100 neural network models that had the best prediction performance as the remaining useful life prediction model for the final milling cutter, and obtained the best results in the competition.

Huimin Chen [19], the runner-up for the competition, was from the Department of Electronic Engineering at the University of New Orleans. By considering the different wear trends of each milling cutter, Chen proposed a method of remaining useful life prediction based on multiple model fusion. The detailed data analysis process is shown in Fig. 5.11. Considering that the initial wear state for each milling cutter was different from Das's model for predicting the wear of the cutter teeth, Chen proposed a regression model for predicting the increment of the wear of the cutter teeth in each milling process. Through comparative analysis, it was found that the prediction effect of establishing respective models for different milling cutters was better than that of establishing a single model. During the feature extraction process, a large number of features were extracted from the original data using conventional methods, and the false discovery rate in multiple tests was used to select the appropriate feature set for a regression model. After establishing the incremental regression model for the cutter tooth wear for each milling cutter in the training data, a Bayesian framework was used to fuse multiple models to predict the incremental wear for the test milling cutter. Based on the assumption that part of the standard variance of the acoustic emission signal in a high frequency band is related to the initial wear state of the milling cutter, the initial wear state of the milling cutter was estimated by using the acoustic emission signals from 15 milling processes before the milling cutter was tested. Combined with the incremental wear model of the milling cutter, Chen predicted and tested the wear on the milling cutter teeth after each milling process, and achieved exceptional results in the competition.

5.5 Open Source Industrial Big Data Competitions

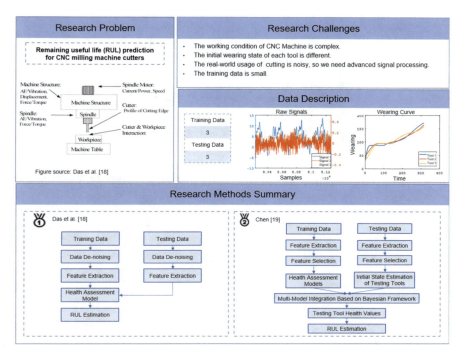

Fig. 5.11 PHM 2010 data competition—RUL of CNC machine tool projects

Although the data analysis processes were similar, compared to the multiple model fusion method proposed by the runner-up, the method proposed by the winning team was simpler and more direct. Because of the effective noise reduction and feature extraction methods for milling characteristics, their models achieved better prediction accuracy.

2011 PHM Data Challenge: Anemometer Health Assessment

A cup anemometer is the most commonly used piece of wind measuring equipment for wind resource assessment, and wind resource assessment is the first step in building a wind power farm. The induction part of the cup anemometer consists of three or four cups. The measurement results are typically used to estimate the future energy output of the site selection. The accuracy of the measurement results will affect the calculation results of the net energy output, and even the assessment of the investment risk, for the wind farm. When the wind speed is about 10 meters per second, a 2% measurement error will lead to 6% difference in energy output calculation, which will affect the calculation for the return on investment (Fig. 5.12).

The PHM data competition in 2011 [20] focused on anemometer health assessment. Data was collected from meteorological towers with altitudes of 50–60 m. Each tower contained several cup anemometers, a wind direction meter, and a temperature sensor. After measuring the data for a 10 min interval, each sensor returned

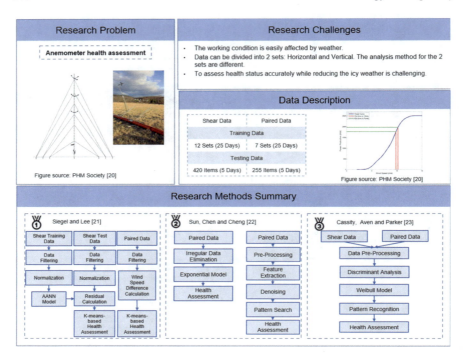

Fig. 5.12 PHM data competition 2011—anemometer health assessment

the mean, standard deviation, minimum, and maximum for the data. On a 60-m-high meteorological tower, there were two anemometers at the heights of 10 and 30 m, and two anemometers at the height of 49 and 59 m; each pair of anemometers was at an angle of 90°. The competition data were divided into two groups: the first consisted of horizontal data, including wind speed data recorded by two anemometers at the same height of the meteorological tower, as well as wind direction and temperature data; the other was vertical data, including wind speed data recorded by anemometers at different heights from the meteorological tower, as well as wind direction and temperature data. Each group of data included multiple training sets with 25-day data and test sets with 5-day data.

The topic of this competition was to test whether there was any degradation of anemometer measurement performance throughout the timeframe for which data was recorded, such as an increase in measurement error caused by bearing wear, transmission shaft failure, or the loss of a cup. The difficulty for the study was that the operating conditions for anemometers is susceptible to the weather, and health monitoring is difficult. Especially in winter weather, anemometers can run slowly due to icing at the physical connection point. In addition, due to the large differences between the horizontal and vertical data sets, it is difficult to select suitable analysis methods, which increases the difficulty of data analysis in general.

5.5 Open Source Industrial Big Data Competitions

The team that won the championship was composed of David Siegel and others from the University of Cincinnati IMS Center, who mainly used auto-associative residual processing and a k-means clustering algorithm [21]. Because of the different characteristics of the two sets of data, the winning team used different methods to analyze them. For horizontal data, firstly, outliers were removed through data filtering and data preprocessing, and erroneous data affected by snow and ice weather were excluded. Then, the difference for the mean data from a pair of anemometers at the same altitude was calculated, and the k-means algorithm was input for clustering analysis. A k-means clustering algorithm divides the difference in the data into two categories and then calculates the mean values of the two types of data separately. The smaller mean values are taken as the calculation results. If the calculation results are less than the threshold value, the anemometer can be considered to be in an unhealthy state.

For vertical data, because the data came from anemometers of different heights, normalization was needed after data filtering and data preprocessing. Afterwards, a complex model was built by using the auto-associative neural network (AANN) algorithm, and the weights of different models were calculated by training data. Then the test data were substituted into the algorithm model to calculate the residual and multiplied by the corresponding weights. The results were divided into two categories by k-means clustering algorithm, and the average values of the two types of data were then calculated separately to get smaller ones. Finally, the mean and threshold values were compared for a health assessment.

The runner-up and second runner-up teams were respectively from Oklahoma State University and Ducommun Incorporated, which is an American military enterprise. The second place team used a similar pattern recognition algorithm based on feature extraction [22]: for horizontal data, they first removed the special values from the data, took the wind speed difference and wind direction data as the core features, and then compared the core features of the test data and training data to conduct a health assessment; for vertical data, they carried out a health assessment first. After data pretreatment and normalization, the power law model was used to fit the data curve. After calculating the sum of residual squares, the health assessment was carried out by comparing the sum of residual squares between test data and training data. The third-place team used the Weibull distribution and discriminant analysis algorithm [23]. This method only selected the horizontal data and a pair of anemometers at the same altitude contained in the vertical data that were similar to the horizontal data, ignoring the data of single anemometers at other altitudes contained in the vertical data. Then, a Weibull distribution model was established for the difference in the data for a pair of anemometers and the parameters were calculated. Finally, the data was classified by a discriminant analysis algorithm to judge the health status of the data.

The championship team's method obtained ideal anemometer health assessment results in the competition, but its processing of horizontal data needed to assume that at least one anemometer pair was in a healthy state in order to give full play to the advantages of the algorithm. Therefore, the follow-up research should consider how to realize the task of anemometer health assessment in actual application scenarios.

2012 IEEE PHM Data Challenge: Remaining Useful Life Estimation of Ball Bearings

The theme of the IEEE PHM 2012 Data Challenge Competition was to estimate the remaining useful life of bearings. Bearings are a widely used key component in manufacturing, power, transportation, and other industries. Most of the failures of rotating machines are related to the degradation and failure of their bearings. The remaining useful life estimation of ball bearings has always been a hot issue in the field of PHM. It is of great significance to improve the availability, safety, and cost-effectiveness of machinery (Fig. 5.13).

The data set for the IEEE PHM 2012 Data Challenge came from the Franche-Comté Électronique Mécanique Thermique et Optique—Sciences et Technologies (FEMTO-ST) Institute [24]. The experiment provided vibration and temperature data under accelerated testing of three different loads (rotating speed and load force). In this competition, six run-to-failure datasets were provided to build a diagnostic model. At the same time, part of the test data of 11 test bearings were provided, which required participants to accurately estimate the effective remaining useful life of these 11 bearings. However, the data did not provide information about the type of bearing failure [25].

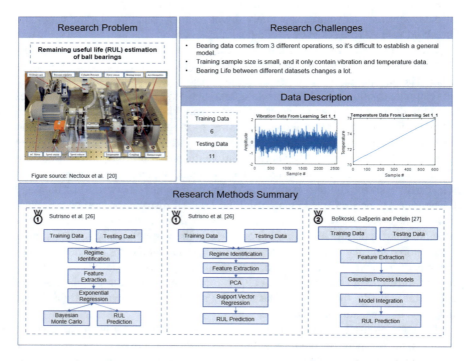

Fig. 5.13 PHM data competition 2012—remaining useful life estimation of ball bearings

The challenges for this particular data competition were due to the following factors: (1) a relative lack of training data; (2) significant variation in experiment duration from one to seven hours; and (3) experimental observations differed greatly from theoretical analysis.

In this data challenge, the winning team [26] considered many different data-driven modeling methods, including moving average spectral kurtosis, Bayesian Monte Carlo sampling, support vector regression, and so on. In the first stage of data analysis, the team found that the commonly used time-domain statistical features and frequency-domain features could not reflect the stable bearing degradation trend, so they turned to study how to model signal processing based on the time-domain and frequency-domain features. After moving average filtering, the upward trend of kurtosis values was identified. Furthermore, based on correlation coefficient analysis, the team found that the kurtosis characteristics of vibration signals after 5.5–6.0 kHz bandpass filtering could well reflect the declining trend of the bearings. In the modeling stage, firstly, the exponential fitting was carried out based on the selected moving average kurtosis signal, and then the remaining useful life and its distribution of bearings were determined based on Bayesian Monte Carlo sampling. The model increased the number of features and improved the shortcomings of having less training data. Bayesian Monte Carlo sampling provided the probability distribution and prediction results of RUL.

In the modeling method based on support vector regression, 34 sets of features were extracted from vibration signals by high-order analysis and wavelet transformation, and then normalized and smoothed. After principal component analysis, it was found that the first three principal components already contained 99.5% of the data variation. Then, the regression model between the first three principal components and the remaining useful life of the bearing was established using support vector regression. The model captured dynamic information effectively by time-frequency analysis, and avoided the risk of over-fitting by principal component analysis. Besides, other regression algorithms, such as Gaussian process models [27], also have potentials to achieve good results.

2014 PHM Data Challenge: Asset Risk Assessment in Industrial Remote Monitoring

The PHM competition in 2014 [28] focused on the common problem of asset health calculation in industrial remote monitoring. Participants needed to calculate the health value of the equipment and then classify the equipment into high-risk or low-risk categories according to the health value. For a device, three days or less until failure was defined as high-risk, and outside of three days until failure was defined as low-risk (Fig. 5.14).

Competition data were divided into a training set and a test set, which had 9199 and 9897 samples, respectively. The training set contained three types of data: a parts maintenance record, an equipment use record, and a failure record. The parts maintenance record included equipment numbers, time schedules, maintenance reason numbers, and replacement part numbers. The equipment usage record included equipment numbers, time schedules, and usage logs. The failure record gave the equipment numbers and failure times. The test set contained three types of data. The

150 5 How to Establish Industrial AI Technology and Capability

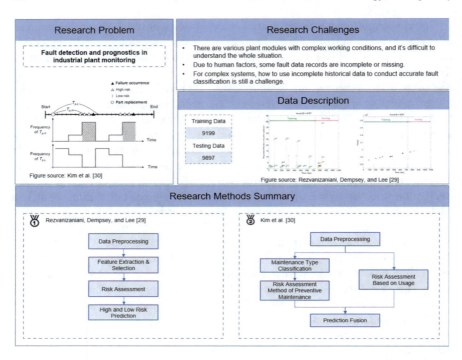

Fig. 5.14 PHM data competition 2014—asset risk assessment in industrial remote monitoring

parts maintenance record and the equipment usage record were the same as the training set, but the contents of the failure record were changed to reflect the equipment number and test time. It was required to evaluate the health value of the equipment during the test time and to assess whether the equipment was at high- or low-risk. For any given device, three years of daily data logs were provided with the first two years serving as training samples, and the third year as the testing sample. Figure 5.16 shows the frequency of failure of a single device, the number of replacement units per failure, and the cause of failure. It also shows the trends of equipment usage increasing over time. The evaluation results of the participants needed to include the health value of the evaluation and the threshold of dividing the high-risk and low-risk categories. The number of high-risk and low-risk samples in the test data were the same. The results of the participants' scores were the sum of the rates correctly assessed as either high-risk or low-risk, respectively.

This challenge presented the following difficulties: first, the specific information of the equipment was not given, so it was impossible to use professional knowledge for analysis. Also, maintenance information rules were complex and needed in-depth excavation. At the same time, equipment usage also had an impact on risk. It may have been necessary to integrate maintenance information and usage information to give a comprehensive evaluation. In addition, the training data contained many outliers, which interfered with the results.

The championship team from this competition was made up of Rezvanizaniani and others from the IMS Center of the University of Cincinnati [29], who used the probabilistic risk assessment method. First, they removed the outliers of usage through data preprocessing. Then, features were found to identify "preventive maintenance"; preventive maintenance here refers to the maintenance of equipment in accordance with a certain period of time. Specifically, by counting the number of replacement parts for each piece of equipment, they were able to find the points in time when the number of parts being replaced was high and assess that these periods had occurred at stable intervals. Having identified cycles of preventative maintenance, maintenance events that occurred during the remaining time could then be regarded as "repair maintenance". For preventive maintenance, in each average maintenance interval, the time interval of high-risk failure was counted. For repair maintenance, according to the bathtub curve, a period of time occuring after repair was a high-risk period. The above two kinds of maintenance define two high-risk periods, respectively. If the test sample failed in any period, it was considered to be high-risk.

The runner-up team for the competition was made up of Hyunjae Kim and others from Seoul National University [30]. They adopted a risk assessment method that integrated maintenance and usage information. First, the method calculated the ratio of high-risk maintenance in the training data by randomly selecting a sample from within the entire data set, and the probability of high-risk maintenance was assessed to be 2.7%. Then, based on the maintenance information, if maintenance occurred within three time points before the failure, it was considered as high-risk maintenance, otherwise it was low-risk maintenance. Taking a certain maintenance action as the starting point, the ratio of high-risk maintenance to total maintenance in each time interval was considered to be a high-risk period if it was greater than 2.7%. The risk assessment method based on usage divided the usage into N intervals and calculated the ratio of high-risk maintenance in each interval. If the ratio was greater than 2.7%, it was considered as a high-risk period. Finally, the evaluation results of the two methods were integrated. The method was that any sample, as long as it fell into any high-risk period for the above two methods, was considered to be high-risk—otherwise, it was considered low-risk.

Fault prediction results for this competition were able to predict some faults more accurately, but the accuracy of the prediction still has room for substantial improvement. After the competition, the champion team put forward the following suggestions to improve the prediction results: make full use of the usage information to locate what is preventive maintenance, utilize the usage information to cluster the data into different categories for analysis, and further excavate the relationship between the parts replacement code and the parts replacement number. The champion team believed that the structure, load, and physical failure information of the equipment should be fused on the basis of current methods, so as to improve the prediction effect and make it easier for industrial applications.

2015 PHM Data Challenge: Fault Detection and Prognostics in Industrial Plant Monitoring

The PHM data competition in 2015 focused on power plant operational failures. Competitors were challenged to predict the type, as well as the start and end times, of a number of plant operational failures. The competition data analysis training set, validation set, and test set featured 33, 15, and 15 samples, respectively. Each factory consisted of different areas, and each area had different components. Moreover, different factories had different regions and numbers of components. The data set contained four types of data: component-based sensor data, control signals, region-based sensor data, and factory fault records. The content categories for the fault records were fault type, as well as start and end times for the fault. There were six fault types, but only five of them required attention during the competition. The sampling interval for the data was 15 min, and the time span for the whole data set was 3–4 years.

As complex systems, power plants contain a wide variety of components, and actual operational conditions vary, so it was difficult to accurately describe the operational status of the power plant in the data set. On top of that, different factories have different working mechanisms and states, so it was difficult to develop a generalized global model. Finally, the physical information given in this competition was limited, and it was impossible to make further use of professional background to do feature engineering.

The first-place team from this competition was Wei Xiao from SAS Institute Inc [31], who used probability prediction in machine learning, as shown in Fig. 5.15. To begin, a significant amount of data was analyzed and visualized, and variable correlation on a time scale was introduced for feature extraction. Secondly, cross validation was designed to ensure the generalization ability of the model. Finally, the start probability for each fault type was predicted at each time period and compared with the threshold value; the most probable start times (no more than two) and end times (no more than two) were selected. Considering the correlation between variables and the non-linearity of the system, machine learning models like k-nearest neighbor, naive Bayes, random forest, gradient boosting machine, and penalized logistic regression were ensembled to output the final results. Meanwhile, the team with the second-highest score from Seoul National University [32] interpreted the physical meaning of the data to select proper features, applied Fisher discriminant analysis to minimize the influence of outliers, and finally recovered the type of missing fault logs and the duration of the faults.

The competition achieved good prediction accuracy, which shows the potential for data mining in complex systems engineering. However, there is still some potential to be tapped, such as clustering power plant samples, modeling, and analysis of each group of similar power plants to further ensure the generalization ability of the model. In addition, the training data in this competition was provided mostly in the form of a time series, and the original sample input data were in matrix form. Thus, the convolution structure of a deep neural network can be considered to learn features, classify, and then make predictions.

2016 PHM Data Challenge: Virtual Metrology of Chemical Mechanical Planarization for Semiconductor Manufacturing

Chemical mechanical planarization (CMP) [33] is an important process in wafer manufacturing, which is used to polish the wafer surface. The planarization process in CMP uses a technology that fixes silicon dioxide, polycrystalline silicon, or a

5.5 Open Source Industrial Big Data Competitions

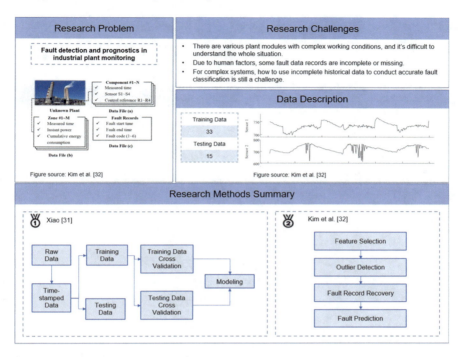

Fig. 5.15 PHM data competition 2015—fault detection and prognostics in industrial plant monitoring

metal layer onto a polishing pad and processes it with a corrosive chemical abrasive fluid. Chemical processes in CMP include passivation and etching of wafer materials with abrasive fluid, while mechanical processes in CMP use downward force to enhance the chemical reaction of etching during the movement of the wafer surface relative to the abrasive fluid particles. CMP is widely used in semiconductor manufacturing because mechanical abrasion alone may cause damage to the surface of materials, and chemical etching alone cannot achieve a good polishing effect. Therefore, the polishing process is completed using the dual effects of physical abrasion and chemical corrosion.

As shown in Fig. 5.16, typical CMP devices include a rotary table, a replaceable polishing pad, a rotating wafer bearing plate, and a rotary dresser. The wafer to be polished is fixed on the lining of the carrying disc; thus, the buckled environmental protection certified wafer on the carrying disc is always in the correct horizontal position. During the polishing process, the polishing pad rotates with the bearing disc, and the downward force acting on the bearing disc rests the wafer on the polishing pad. The grinding fluid from the grinding fluid distributor consists of corrosive particles and other chemicals. The dresser in Fig. 5.16 is made of a hard material, such as diamond, to increase the roughness of the pad surface.

The 2016 PHM Data Challenge focused on the typical CMP process with the goal of predicting the average material removal rate (MRR) of wafer materials. During the polishing process, the flattening ability of the planarization liner decreases with

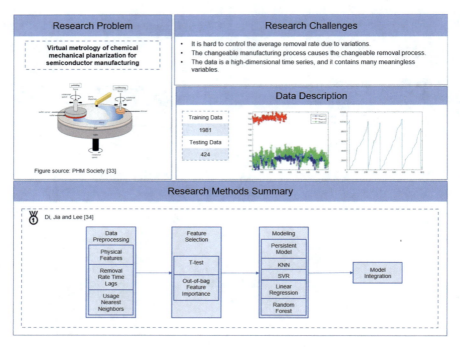

Fig. 5.16 2016 PHM data competition—virtual metrology of CMP for semiconductor manufacturing

time. Therefore, after every processing period, the polishing pad must be replaced. Similarly, the roughening ability of the dresser on the liner will also decrease, and after a series of dressing work, the dresser must be replaced. The competition provided state data for the polishing pads, dressers, and other components. The participants combined physical models to analyze the data to predict MRR; the predicted MRR feedback controller was used to optimize the control parameters. By optimizing the pressure, flow rate, and speed of the CMP process, the process was adapted to the change of processing performance caused by material degradation. The prediction results were measured by mean square error (MSE).

During the CMP monitoring process, the following four kinds of process variables are usually collected:

1. Usage variables for consumable materials, such as linings, polishing pads, and dressers
2. Pressure signals, such as pressure in the processing chamber and buckle pressure, etc.
3. Flow rate of grinding liquefied chemicals
4. Rotation speed of the wafer, polishing pad, and dresser

5.5 Open Source Industrial Big Data Competitions

All these variables have more or less the same influence on the performance of CMP devices, namely MRR. Because the data from the CMP process involves high-dimensional time variables and the working conditions are changeable, it is difficult to control the consumption rate for the wafer materials. In order to describe the CMP process in as much detail as possible, we must use as many variables as possible. In data analysis, if we cannot determine which variables play a decisive role in the final MRR, and which variables are of little significance, the results of the data analysis will be greatly discounted.

The winning team [34] for the 2016 Championship used the physical meaning of variables to screen all the data. Because they were able to extract high-quality features, the best prediction results were obtained. The method first classified the data according to the different processing chambers and stages.

The feature extraction part was divided into physical features, time neighbor of the loss rate, and material consumption neighbor. In this research method, physical characteristics refer to the statistics of each relevant physical variable including mean, standard deviation, peak-to-peak value, and so on. In addition, because CMP is a continuous process, the change of MRR can also be regarded as a time series, so this method extracted the MRR for the nearest time—that is, the time neighbor of the removal rate—as one of the features. On the other hand, in the same processing environment, when the usage of consumable materials such as liners, polishing pads, and dressers is the same, the MRR of wafers should be similar. Therefore, the MRR of wafers with similar material usage was extracted as the last set of features.

During the feature selection part, classical t-test and OOB (Out-Of-Bag) were used as criteria to determine the importance of features. All features extracted were further screened in order to retain features with large amounts of information and remove features that had little effect on predicting MRR. During the model building phase, the selected features were used as input, and a learning model based on an intelligent algorithm was established to predict MRR. Five basic learning models were used for this method, and the prediction results were obtained by cross validation and adjusting the weights of each model. In this way, the advantages of each model could be synthetically utilized, and the strengths could be reinforced while the weaknesses were avoided.

The prediction method adopted by the 2016 championship team considered the physical significance of variables, extracting high-quality features, and building a comprehensive model. Finally, the MSE of the prediction results was reduced to a very ideal range, which proves that this method can help improve the control accuracy of CMP processing and help wafer manufacturers to improve their competitiveness.

2017 PHM Data Challenge: Fault Detection and Diagnosis of Train Bogies

The PHM competition in 2017 [35] focused on fault detection and diagnosis in train bogies. The competition focused on the combination of physical models and statistical models. Through the integration of the models, participants established the mapping between the health status of each component of the train bogie, model parameters, and data.

The traditional train bogie system consists of a body, two steering wheels, and four-wheel sets, as shown in the schematic diagram of the bogie vehicle model in Fig. 5.17. The simplified model includes the helical spring and damper in the main suspension and the air spring in the secondary suspension. Sensors are placed on wheels, bogie frames, and the body. Irregular rough tracks and component failures cause vibration in the direction of each axis. In the design experiment, the vehicle ran at different speeds on different pavements, different layouts, and different geometric contact surfaces. In addition, the vehicle parameters such as load, stiffness, and damping rate varied within a certain range, even in the control group without component failure.

The first task in the competition was to use experimental data and physical models to predict whether there had been a fault during train operation. The second task was to locate the fault. The data provided by the competition were features extracted from the original sensor data in the frequency domain, which were divided into training data sets and test data sets, each of which had 200 sets of samples in different experimental states. Each group of samples consisted of 90 features. The training data set only contained the data under the normal operational conditions for each part of the train bogie—that is, the data under the healthy state of the system—while the test data included the data under the healthy state and the fault state. The score for this competition was based on the sum of the scores of the two tasks. The first objective

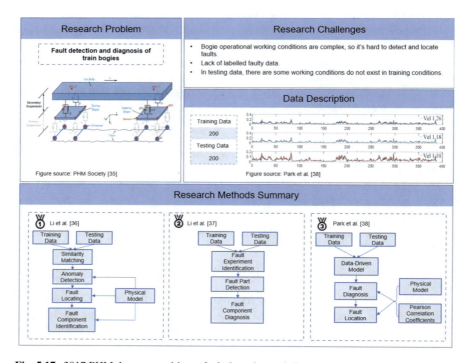

Fig. 5.17 2017 PHM data competition—fault detection and diagnosis of train bogies

5.5 Open Source Industrial Big Data Competitions

was to identify health and faults accurately, and the second was to locate faults by sensitivity, which divided the number of faulty components correctly predicted by the total number of faults.

The challenges for this competition mainly came from the following aspects: the complex running environment of the train and various working conditions affected the fault diagnosis, including the lack of data in the fault state and the incomplete experimental conditions of training data. These affected the performance of the trained model in unknown states. The insufficient amount of evidence from data and training samples may have led to over-fitting and other problems.

The champions of this competition were Chuang Li and others from K2Data Technology in China [36], who adopted a hybrid model based on similarity. First, they used a similarity matching method to preprocess data to reduce the impact of track irregularities. Then, 31 similar/related features were extracted from the physical model to correspond to the abnormal offset/fade in the machine learning model. Next, the construction mapping matrix between position and feature was summarized from the physical model. Finally, the physical model was used to locate the damper fault, spring fault, or both.

The second-place team in this competition was Sanhua Li and others from the Innovation Center for Industrial Big Data in China [37], who used the ensemble model to diagnose faults. First, the ensemble model, which is based on a linear model, was used to judge whether the experimental sample was in a fault state; it extracted the factors of vibration attenuation in the system, extracted the sequence features from the data by sliding average, and then modeled the external factors. Second, a fault detection method based on residual analysis and pattern mining was proposed. Finally, the fault mode template extracted from the physical model was applied to the machine learning model to determine the fault types.

The third-place team was Chan Hee Park and others from Seoul National University in Korea [38], who used a data-driven and model-driven hybrid model. First, the root mean square error for the training set and test set was calculated by the data-driven method, and the maximum root mean square error could then be used to identify the faults for the components near the sensor. Next, an ensemble model based on the physical model and the Pearson correlation coefficient model was used to judge the data in unknown orbits. In the physical model, the maximum root mean square error could be used to identify the faults for the components near the sensor. For each suspension, a transfer function was designed to connect the relevant sensors. Using the Pearson correlation coefficient, independent correlation coefficient values relative to road conditions were used to detect and identify faults.

Based on the results of this competition, it is clear that the combination of the physical and data-driven models were able to achieve better results in identifying the main task, i.e. fault pattern. In particular, a physical model can be applied to fault identification in the absence of data and system background information. For the task of fault location in this competition, the winning teams all hoped to use the mechanism of a physical model to judge the fault location, but there is still room for substantial improvement given the current results. The idea of combining a physical model with a data model is an important research trend in recent years. While a data-driven model is helpful by having good practicality and extensibility, with the help of the physical model, the "black box" of the data-driven model can be opened, and the mechanism knowledge of the system can be better transformed into useful information during the process of modeling. How to better integrate the two and have them learn from each other's strengths and weaknesses is also one of the future trends in the field of PHM.

References

1. De Carolis A, Macchi M, Negri E, Terzi S (2017, September) A maturity model for assessing the digital readiness of manufacturing companies. In: IFIP international conference on advances in production management systems. Springer, Cham, pp 13–20
2. Jung K, Kulvatunyou B, Choi S, Brundage MP (2016, September) An overview of a smart manufacturing system readiness assessment. In: IFIP international conference on advances in production management systems. Springer, Cham, pp 705–712
3. MESA (2011) Transforming manufacturing maturity with ISA-95 methods
4. EDB Singapore (2019) The Singapore smart industry readiness index. Available: https://www.edb.gov.sg/en/news-and-events/news/advanced-manufacturing-release.html
5. WEF (2019) World economic forum. Available: https://www.weforum.org/
6. WEF (2019) Fourth industrial revolution beacons of technology and innovation in manufacturing. Available: http://www3.weforum.org/docs/WEF_4IR_Beacons_of_Technology_and_Innovation_in_Manufacturing_report_2019.pdf
7. Osawa Y, Miyazaki K (2006) An empirical analysis of the valley of death: large-scale R&D project performance in a Japanese diversified company. Asian J Technol Innov 14(2):93–116
8. Frederick DK, DeCastro JA, Litt JS (2007) User's guide for the commercial modular aero-propulsion system simulation (C-MAPSS). Available: https://ntrs.nasa.gov/archive/nasa/casi.ntrs.nasa.gov/20070034949.pdf
9. NASA (2019) PCoE datasets. Available: https://ti.arc.nasa.gov/tech/dash/groups/pcoe/prognostic-data-repository/
10. Wang T, Yu J, Siegel D, Lee J (2008) A similarity-based prognostics approach for remaining useful life estimation of engineered systems. In: Proceeding of 2008 international conference on prognostics and health management. IEEE
11. Peel L (2008) Data driven prognostics using a Kalman Filter ensemble of neural network models. In: Proceeding of 2008 international conference on prognostics and health management. IEEE
12. Heimes FO (2008) Recurrent neural networks for remaining useful life estimation. In: Proceeding of 2008 international conference on prognostics and health management. IEEE
13. PHM society (2009) 2009 PHM challenge competition data set. Available: https://www.phmsociety.org/references/datasets
14. PHM society (2009) Apparatus for the PHM09 data challenge. Available: https://www.phmsociety.org/competition/PHM/09/apparatus
15. Wu F, Lee J (2011) Information reconstruction method for improved clustering and diagnosis of generic gearbox signals. Int J Progn Health Manag 2:42
16. Al-Atat H, Siegel D, Lee J (2011) A systematic methodology for gearbox health assessment and fault classification. Int J Progn Health Manag 2(1):16
17. Boškoski P, Urevc A (2011) Bearing fault detection with application to PHM data challenge. Int J Progn Health Manag 2(color):32
18. Das S, Hall R, Herzog S, Harrison G, Bodkin M, Martin L (2011) Essential steps in prognostic health management. In: Proceeding of 2011 IEEE conference on prognostics and health management. IEEE, pp 1–9

References

19. Chen H (2011) A multiple model prediction algorithm for CNC machine wear PHM. Int J Progn Health Manag 2:129
20. PHM Society (2011) PHM data challenge 2011—condition monitoring of anemometers. Available: https://www.phmsociety.org/competition/phm/11/problem
21. Siegel D, Lee J (2011) An auto-associative residual processing and K-means clustering approach for anemometer health assessment. Int J Progn Health Manag 2(color):117
22. Sun L, Chen C, Cheng Q (2012) Feature extraction and pattern identification for anemometer condition diagnosis. Int J Progn Health Manag 3:8–18
23. Cassity J, Aven C, Parker D (2012) Applying Weibull distribution and discriminant function techniques to predict damaged cup anemometers in 2011 PHM competition. Int J Progn Health Manag 3:1–7
24. FEMTO-ST (2012) IEEE PHM 2012 Data challenge. Available: http://www.femto-st.fr/en/Research-departments/AS2M/Research-groups/PHM/IEEE-PHM-2012-Data-challenge.php
25. Nectoux P, Gouriveau R, Medjaher K, Ramasso E, Chebel-Morello B, Zerhouni N, Varnier C (2012) PRONOSTIA: an experimental platform for bearings accelerated life test. In: IEEE international conference on prognostics and health management. Denver, CO, USA. Available from: https://www.scribd.com/document/344094074/IEEEPHM2012-Challenge-Details-Results-and-Winners
26. Sutrisno E, Oh H, Vasan ASS, Pecht M (2012) Estimation of remaining useful life of ball bearings using data driven methodologies. In: 2012 IEEE conference on prognostics and health management. IEEE, pp 1–7
27. Boškoski P, Gašperin M, Petelin, D (2012, June) Bearing fault prognostics based on signal complexity and Gaussian process models. In: 2012 IEEE Conference on Prognostics and Health Management IEEE, pp 1–8
28. PHM Society (2014) PHM data challenge 2014. Available: https://www.phmsociety.org/events/conference/phm/14/data-challenge
29. Rezvanizaniani SM, Dempsey J, Lee J (2014) An effective predictive maintenance approach based on historical maintenance data using a probabilistic risk assessment: PHM14 data challenge. Int J Progn Health Manag 5(2)
30. Kim H, Hwang T, Park J, Oh H, Youn BD (2014) Risk prediction of engineering assets: an ensemble of part lifespan calculation and usage classification methods. Int J Progn Health Manag 5(2)
31. Xiao W (2016) A probabilistic machine learning approach to detect industrial plant faults. Int J Progn Health Manag
32. Kim J, Ha JM, Park J, Kim S, Kim K, Jang BC, … Youn BD (2016) Fault log recovery using an incomplete-data-trained FDA classifier for failure diagnosis of engineered systems. Int J Progn Health Manag
33. PHM Society (2016) 2016 PHM Data Challenge. Available: https://www.phmsociety.org/events/conference/phm/16/data-challenge
34. Di Y, Jia X, Lee, J. (2017). Enhanced virtual metrology on chemical mechanical planarization process using an integrated model and data-driven approach. Int J Progn Health Manag 8(2)
35. PHM Society (2017) 2017 PHM Data Challenge. Available: https://www.phmsociety.org/events/conference/phm/17/data-challenge
36. Li C, Liu J, Tian C, Cui P, Wu M (2017) Similarity-based fault detection in vehicle suspension system. In: Annual conference of the prognostics and health management society 2017
37. Li S, Tian Y, Jing Z, Huang Y, Yang Y (2017) Ensemble model based fault prognostic method for railway vehicles suspension system. In: Annual conference of the prognostics and health management society 2017
38. Park CH, Kim S, Lee J, Lee D-K, Na K, Song J, Youn BD (2017) Hybriding data-driven and model-based approaches for fault diagnosis of rail vehicle suspensions. In: Annual conference of the prognostics and health management society 2017

Chapter 6
Conclusion

The future direction of Industrial Internet technology is to transform industrial processes and products, moving from the reliance on experience-based decision making towards data-centric or evidence-based decision making. It is through this process that Industrial AI will play a major role to advance the digitalization of traditional industrial systems.

After reading this book, I hope you now understand the difference between Industrial AI and traditional AI methods. At its core, Industrial AI is an intelligent system based on the concrete needs and specific problems of the industrial domain. Therefore, in the last chapter of this book, we reviewed these concepts using real-life research examples. While the impetus of traditional AI stems from the divergent needs of life, society, and finance, often resulting in the creation of novel applications driven by social needs and interests, Industrial AI is a comprehensive and systematic methodology working convergently to improve the reliability, precision, efficiency, and future optimization of engineering systems.

This book also introduced some Industrial AI system engineering methods, how to move from traditional software algorithms and algorithm-centric thinking towards systems integration (i.e. the use of high-performance computing (HPC) and tools), and finally demonstrated the value of Industrial AI with examples of real-life practical applications. Currently, Industrial AI is gradually being applied across industry in areas such as machine monitoring, transportation fuel efficiency, engine health management, oil field and refinery safety and reliability management, and the remote maintenance of health care systems. Despite these advances, however, the concept of system engineering is still lacking, as is the basis for sustainable knowledge transfer. At the same time, this book identified many cases of Industrial AI applications, and in a more intuitive way, it describes the development of Industrial AI technology and shares lessons learned in the process.

As future industry transforms, we will find that Industrial AI will play an important role in how to make up for the experience and knowledge of craftsmen lost in this wave

© Shanghai Jiao Tong University Press 2020
J. Lee, *Industrial AI*,
https://doi.org/10.1007/978-981-15-2144-7_6

of automation. I hope readers have gained a comprehensive understanding of Industrial AI by reading this book, and will proceed to apply the relevant knowledge of Industrial AI to their respective professions to tap into invisible knowledge, to attempt to break through the limitations of traditional experience-based manufacturing, and ultimately to create greater value.